Biogeography and Geomorphology

Biogeography and Geomorphology

J.K. SINHA

Orange Boooks International
NEW DELHI-110002

ISBN : 978-93-83263-73-8

Published in 2018 in India by

Orange Boooks International

Contact Us
113A, Prakasha Scout,
Ansari Road, Daryaganj,
New Delhi - 110 002
INDIA
Phone:- 011-43046297,
011-43522777
Email:- Info@obipd.com

Preface

Biogeography is the study of the distribution of species and ecosystems in geographic space and through geological time. Organisms and biological communities often vary in a regular fashion along geographic gradients of latitude, elevation, isolation and habitat area. Phytogeography is the branch of biogeography that studies the distribution of plants. Zoogeography is the branch that studies distribution of animals.

Biogeography deals with the geography, ecology, and history of life where it lives, how it lives there, and how it came to live there. It has three main branches analytical biogeography, ecological biogeography, and historical biogeography. Historical biogeography considers the influence of continental drift, global climatic change, and other large-scale environmental factors on the long-term evolution of life. Ecological biogeography looks at the relations between life and the environmental complex. Analytical biogeography examines where organisms live today and how they spread. It may be considered as a division of ecological biogeography.

Soil is one of the three major natural resources, alongside air and water. It is one of the marvellous products of nature and without which there would be no life. Soil is made up of three main components – minerals that come from rocks below or nearby, organic matter which is the remains of plants and animals that use the soil, and the living organisms that reside in the soil. The proportion of each of these is important in determining the type of soil that is present. But other factors such as climate, vegetation, time, the surrounding terrain, and even human activities (eg. farming, grazing, gardening etc.), are also important in influencing how soil is formed and the types of soil that occur in a particular landscape.

Geomorphology is the study of landforms and landscapes, including the description, classification, origin, development, and history of planetar surfaces. During the early part of this century, the study of regional-scale geomorphology was termed "physiography". Unfortunately, physiography also became synonymouswith physical geography, and the concept became embroiled in controversy surrounding the appropriate concerns of that discipline. Some geomorphologists held to a geological basis for physiography and emphasized a concept of physiographic regions, A conflicting trend among geographers was to equate physiography with "pure morphology," divorced of its geological heritage.

The surface of the earth is modified by a combination of surface processes that sculpt landscapes, and geologic processes that cause tectonic uplift and subsidence, and shape the coastal geography. Surface processes comprise the action of water, wind, ice, fire, and living things on the surface of the earth, along with chemical reactions that form soils and alter material properties, the stability and rate of change of topography under the force of gravity, and other factors, such as (in the very recent past) human alteration of the landscape.

Hydrogeology is an interdisciplinary subject; it can be difficult to account fully for thechemical, physical, biological and even legal interactions between soil, water, natureand society. The study of the interaction between groundwater movement and geology can be quite complex. Groundwater does not always flow in the subsurface down-hill following the surface topography; groundwater follows pressure gradients(flow from high pressure to low) often following fractures and conduits in circuitous paths.

This is a reference book. All the matter is just compiled and edited in nature, taken from the various sources which are in public domain.

The book also combines science and policy perspectives to the biogeography and geomorphology. It is useful for students, researchers and teachers coming from geography, environmental studies, biosciences, ecology and policy science.

—*Editor*

Contents

1

Biogeography

INTRODUCTION

Biogeography deals with the geography, ecology, and history of life where it lives, how it lives there, and how it came to live there. It has three main branches analytical biogeography, ecological biogeography, and historical biogeography. Historical biogeography considers the influence of continental drift, global climatic change, and other large-scale environmental factors on the long-term evolution of life. Ecological biogeography looks at the relations between life and the environmental complex. Analytical biogeography examines where organisms live today and how they spread. It may be considered as a division of ecological biogeography.

The ecological and historical biogeography of animals and plants, and, in doing so, it considers human involvement in the living world. It is designed to lead students through the main areas of modern biogeographical investigation. Basic ideas are carefully explained using numerous examples from around the world.

What Is Biogeography?

Biogeographers study the geography, ecology, and evolution of living things.

- Ecology—environmental constraints on living
- History and geography—time and space constraints on living

Biogeographers address a misleadingly simple question: why do organisms live where they do? Why is the speckled rangeland grasshopper confined to short-grass prairie and forest or brushland clearings containing small patches of bare ground? Why does the ring ouzel live in Norway, Sweden, the British Isles, and mountainous parts of central Europe, Turkey, and south-west Asia, but not in the intervening regions? Why do tapirs live only in South America and South-east Asia? Why do the nestor parrots—the kea and the kaka—live only in New Zealand? Why do pouched mammals (marsupials) live in Australia and the Americas, but not in Europe, Asia, Africa, or Antarctica? Why do different regions carry distinct assemblages of animals and plants? Two groups of reasons are given in answer to such questions as these—ecological reasons and historical-cum-geographical reasons.

Ecology

Ecological explanations for the distribution of organisms involve several interrelated ideas. First is the idea of populations, which is the subject of analytical biogeography. Each species has a characteristic life history, reproduction rate, behaviour, means of dispersion, and so on. These traits affect a population's response to the environment in which it lives. The second idea concerns this biological response to the environment and is the subject of ecological biogeography. A population responds to its physical surroundings (abiotic environment) and its living surroundings (biotic environment). Factors in the abiotic environment include such physical factors as temperature, light, soil, geology, topography, fire, water, water and air currents; and such chemical factors as oxygen levels, salt concentrations, the presence of toxins, and acidity. Factors in the biotic environment include competing species, parasites, diseases, predators, and humans. In short, each species can tolerate a range of environmental factors. It can only live where these factors lie within its tolerance limits.

Speckled Rangeland Grasshopper

This insect *(Arphia conspersa)* ranges from Alaska and northern Canada to northern Mexico, and from California to the Great Plains. It is found at less than 1,000 m elevation in the northern part of its range and up to 4,000 m in the southern part. Within this extensive latitudinal and altitudinal range, its istribution pattern is very patchy, owing to its decided preference for very specific habitats. It requires short-grass prairie, or forest and brushland openings, peppered with small pockets of bare ground. Narrow-leaved grasses provide the grasshopper's food source. The bare patches are needed for it to perform courtship rituals. These ecological and behavioural needs are not met by dense forest, tall grass meadows, or dry scrubland. Roadside meadows and old logged areas are suitable and are slowly being colonized. Moderately grazed pastures are also suitable and support large populations.

Even within suitable habitat, the grasshopper's distribution is limited by its low vagility. This is the result of complex social behaviour, rather than an inability to fly well. Females are fairly sedentary, at least in mountain areas, while males make mainly short, spontaneous flights within a limited area. The two sexes together form tightly knit population clusters within areas of suitable habitat. The clusters are held together by visual and acoustic communication displays.

Ring Ouzel

The biogeography of most species may be explained by a mix of ecology and history. The ring ouzel or 'mountain blackbird', which goes by the undignified scientific name of *Turdus torquatus*, lives in the cool temperate climatic zone, and in the alpine equivalent to the cool temperate zone on mountains. It likes cold climates. During the last ice age, the heart of its range was probably the Alps and Balkans. From here, it spread outwards into much of Europe, which was then colder than now. With climatic warming during the last 10,000 years, the ring ouzel has left much of its former range and survives only in places that are still relatively cold because of their high latitude or altitude. Even though it likes cold conditions, most ring ouzels migrate to less severe climates during winter. The north European populations move to the Mediterranean while the alpine populations move to lower altitudes.

HISTORY AND GEOGRAPHY

Historical-cum-geographical explanations for the distribution of organisms involve two basic ideas, both of which are the subject of historical biogeography. The first idea concerns centres-of-origin and dispersal from one place to another. It argues that species originate in a particular place and then spread to other parts of the globe, if they should be able and willing to do so. The second idea considers the importance of geological and climatic changes splitting a single population into two or more isolated groups. This idea is known as vicariance biogeography. These two basic biogeographical processes are seen in the following case studies.

Tapirs

The tapirs are close relatives of the horses and rhinoceroses. They form a family—the Tapiridae. There are four living species, one of which dwells in Southeast Asia and three in Central and South America. Their present distribution is thus broken and poses a problem for biogeographers. How do such closely related species come to live in geographically distant parts of the world? Finds of fossil tapirs help to answer this puzzle. Members of the tapir family were once far more widely distributed than at present. They are known to have lived in North America and Eurasia. The oldest fossils come from Europe. A logical conclusion is that the tapirs evolved in Europe, which was their centre of origin, and then dispersed east and west. The tapirs that went north-east reached North America and South America. The tapirs that chose a south-easterly dispersal route moved into South-east Asia. Subsequently, probably owing to climatic change, the tapirs in North America and the Eurasian homeland went extinct. The survivors at the tropical edges of the distribution spawned the present species. This explanation is plausible, though it is not water tight it is always possible that somebody will dig up even older tapir remains from somewhere else. Historical biogeographers are dogged by the incompleteness of the fossil record, which means that they can never be fully confident about any hypothesis.

LIFE AND THE ENVIRONMENT

Life is adapted to nearly all Earth surface environments:

- Places to live
- Climatic constraints on living
- Other physical constraints on living
- Ways of living

Individuals, species, and populations, both marine and terrestrial, tend to live in particular places. These places are called habitats. Each habitat is characterized by a specific set of environmental conditions—radiation and light, temperature, moisture, wind, fire frequency and intensity, gravity, salinity, currents, topography, soil, substrate, geomorphology, human disturbance, and so forth.

A place to live: habitats

Habitats come in all shapes and sizes, occupying the full sweep of geographical scales. They range from small (microhabitats), through medium (mesohabitats) and large (macrohabitats), to very large (megahabitats). Microhabitats are a few square centimetres to a few square metres in area. They include leaves, the soil, lake bottoms, sandy beaches, talus slopes, walls, river banks, and paths. Mesohabitats have areas up to about 10,000 km^2; that is, a 100×100 kilometre square, which is about the size of Cheshire, England. Each main mesohabitat is influenced by the same regional climate, by similar features of geomorphology and soils, and by a similar set of disturbance regimes. Deciduous woodland, caves, and streams are examples. Macrohabitats have areas up to about 1,000,000 km^2, which is about the size of Ireland. Megahabitats are regions more than 1,000,000 km^2 in extent. They include continents and the entire land surface of the Earth. Landscape ecologists, who have an express interest in the geographical dimension of ecosystems, recognize three levels of 'habitat'—region, landscape, and landscape element. These correspond to large-scale, medium-scale, and small-scale habitats. Some landscape ecologists are relaxing their interpretation of a landscape to include smaller and larger scales—they have come to realize that a beetle's view or a bird's view of the landscape is very different from a human's view.

Landscape Elements

Landscape elements are similar to microhabitats, but a little larger. They are fairly uniform pieces of land, no smaller than about 10 m, that form the building blocks of landscapes and regions. They are also called ecotopes, biotopes, geotopes, facies, sites, tesserae, landscape units, landscape cells, and landscape prisms. These terms are roughly equivalent to landscape element, but have their own special meanings. Landscape elements are made of individual trees, shrubs, herbs, and small buildings. There are three basic kinds of landscape element—patches, corridors, and background matrixes:

1. Patches are fairly uniform (homogeneous) areas that differ from their surroundings. Woods, fields, ponds, rock outcrops, and houses are all patches.
2. Corridors are strips of land that differ from the land to either side. They may interconnect to form networks. Roads, hedgerows, and rivers are corridors.
3. Background matrixes are the background ecosystems or land-use types in which patches and corridors are set. Examples are deciduous forest and areas of arable cultivation.

Landscape elements include the results of human toil roads, railways, canals, houses, and so on. Such features dominate the landscape in many parts of the world and form a kind of 'designer mosaic'. Designed patches include urban areas, urban and suburban parks and gardens (greenspaces), fields, cleared land, and reservoirs. Designed corridors include hedgerows, roads and railways, canals, dikes, bridle paths, and footpaths. There is also a variety of undesigned patches—waste tips, derelict land, spoil heaps, and so on.

Landscapes

Landscape elements combine to form landscapes. A landscape is a mosaic, an assortment of patches and corridors set in a matrix, no bigger than about 10,000 km^2. It is 'a heterogeneous land area composed of a cluster of interacting ecosystems that is repeated in similar form throughout'. By way of example, the recurring cluster of interacting ecosystems that feature in the landscape around the author's home, in the foothills of the Pennines, includes woodland, field, hedgerow, pond, brook, canal, roadside, path, quarry, mine tip, disused mining incline, disused railway, farm building, and residential plot.

Regions

Landscapes combine to form regions, more than about 10,000 km^2 in area. They are collections of landscapes sharing the same macroclimate. All Mediterranean landscapes share a seasonal climate characterized by mild, wet winters and hot, droughty summers.

Habitat requirements

It is probably true to say that no two species have exactly the same living requirements. There are two extreme cases—fussy species or habitat specialists and unfussy species or habitat generalists—and all grades of 'fussiness' between.

Habitat Specialists

Habitat specialists have very precise living requirements. In southern England, the red ant, *Myrmica sabuleti,* needs dry heathland with a warm south-facing aspect that contains more than 50 per cent grass species, and that has been disturbed within the previous five years. Other species are less pernickety and thrive over a wider range of environmental conditions. The three-toed woodpecker *(Picoides tridactylus)* lives in a broad swath of cool temperate forest encircling the Northern Hemisphere. Races of the common jay *(Garrulus glandarius)* occupy a belt of oak and mixed deciduous woodland stretching from Britain to Japan.

Habitat Generalists

A few species manage to eke out a living in a great array of environments. The human species *(Homo sapiens)* is the champion habitat generalist — the planet Earth is the human habitat. In the plant kingdom, the broad-leaved plantain *(Plantago major),* typically a species of grassland habitats, is found almost everywhere except Antarctica and the dry parts of North Africa and the Middle East. In the British Isles, it seems indifferent to climate and soil conditions, being found in all grasslands on acid and alkaline soils alike. It also lives on paths, tracks, disturbed habitats (spoil heaps, demolition sites, arable land), pasture and meadows, road verges, river banks, mires, skeletal habitats, and as a weed in lawns and sports fields. In woodland, it is found only in relatively unshaded areas along rides. It is not found in aquatic habitats or tall herb communities.

Life's Limits: Ecological Tolerance

Organisms live in virtually all environments, from the hottest to the coldest, the wettest to the driest, the most acidic to the most alkaline. Understandably, humans tend to think of their 'comfortable' environment as the norm. But moderate conditions are anathema to the micro-organisms that love conditions fatal to other creatures. These are the extremophiles. An example is high-pressure-loving microbes (barophiles) that flourish in deep-sea environments and are adapted to life at high pressures. Many other organisms are adapted to conditions that, by white western human standards, are harsh, though not so extreme as the conditions favoured by the extremophiles. Examples are hot deserts and Arctic and alpine regions.

Limiting Factors

A limiting factor is an environmental factor that slows down population growth. The term was first suggested by Justus von Liebig, a German agricultural chemist. Liebig noticed that the growth of a field crop is limited by whichever nutrient happens to be in short supply. A field of wheat may have ample phosphorus to yield well, but if another nutrient, say nitrogen, should be lacking, then the yield will be reduced. No matter how much extra phosphorus is applied in fertilizer, the lack of nitrogen will limit wheat yield. Only by making good the nitrogen shortage could yields be improved. These observations led to Liebig to establish a 'law of the minimum': the productivity, growth, and reproduction of organisms will be constrained if one or more environmental factors lies below its limiting level.

Later, ecologists established a 'law of the maximum'. This law applies where population growth is curtailed by an environmental factor exceeding an upper limiting level. In a wheat field, too much phosphorus is as harmful as too little—there is an upper limit to nutrient levels that plants can tolerate.

Tolerance Range

For every environmental factor (such as temperature and moisture) there is a lower limit below which a species cannot live, an optimum range in which it thrives, and an upper limit above which it cannot live. The upper and lower bounds define the tolerance range of a species for a particular environmental factor. The bounds vary from species to species. A species will prosper within its optimum range of tolerance; survive but show signs of physiological stress near its tolerance limits; and not survive outside its tolerance range. Stress is a widely used but troublesome idea in ecology. It may be defined as 'external constraints limiting the rates of resource acquisition, growth or reproduction of organisms'.

Each species (or race) has a characteristic tolerance range. Stenoecious species have a wide tolerance; euryoecious species have a narrow tolerance. All species, regardless of their tolerance range, may be adapted to the low end, to the middle, or to the high end (polytypic) of an environmental gradient. Take the example of photosynthesis in plants. Plants adapted to cool temperatures (oligotherms) have photosynthetic optima at about 10°C and cease to photosynthesize above 25°C. Temperate-zone plants (mesotherms) have optima between 15°C and 30°C. Tropical plants (polytherms) may have optima as high as 40°C. Interestingly, these optima are not 'hard and fast'. Cold-adapted plants are able to shift their photosynthetic optima towards higher temperatures when they are grown under warmer conditions.

Ecological Valency

Tolerance may be wide or narrow and the optimum may be at low, middle, or high positions along an environmental gradient. When combined, these contingencies produce six grades of ecological valency. The glacial flea *(Isotoma saltans)* has a narrow temperature tolerance and likes it cold.

CLIMATIC FACTORS

Flower Power: Radiation and Light

The Sun is the primary source of radiation for the Earth. It emits electromagnetic radiation across a broad spectrum, from very short wavelengths to long wavelengths. The visible portion (sunlight) is the effective bit for photosynthesis. It is also significant in heating the environment. Long-wave (infrared) radiation emitted by the Earth is locally important around volcanoes, in geothermal springs, and in hydrothermal vents in the deep-sea floor. These internal sources of energy are tapped by unusual organisms, including the thermophiles and hyperthermophiles that like it very hot.

Three aspects of solar radiation influence photosynthesis—the intensity, the quality, and the photoperiod or duration. The intensity of solar radiation is the amount that falls on a given area in a unit of time. Calories per square centimetre per minute (cal/cm^2/ min) were once popular units, but Watts per square metre (W/m^2) or kiloJoules per hectare (kJ/ha) are metric alternatives. The average annual solar radiation on a horizontal ground surface ranges from about 800 kJ/ha over subtropical deserts to less than 300 kJ/ha in polar regions. Equatorial regions receive less radiation than the subtropics because they are cloudier. A value of 700 kJ/ha is typical.

The quality of solar radiation is its wavelength composition. This varies from place to place depending on the composition of the atmosphere, different components of which filter out different parts of the electromagnetic spectrum. In the tropics, about twice as much ultraviolet light reaches the ground above 2,500 m than at sea level. Indeed, ultraviolet light is stronger in all mountains—hence incautious humans may unexpectedly suffer sunburn at ski resorts.

The Electromagnetic Spectrum Emitted By The Sun

Electromagnetic radiation pours out of the Sun at the speed of light. Extreme ultraviolet radiation with wavelengths in the range 30 to 120 nanometres (nm) occupies the very short end of the spectrum. Ultraviolet light extends to wavelengths of 0.4 micrometres (μm). Visible light has wavelengths in the range 0.4 to 0.8 μm. This is the portion of the electromagnetic spectrum humans can see. Infrared radiation has wavelengths longer than 0.8 μm. It grades into radio frequencies with millimetre to metre wavelengths. The Sun emits most intensely near 5 μm, which is in the green band of the visible light. This fact might help to account for plants being green—they reflect the most intense band of sunlight.

night-length, stimulates the timing of daily and seasonal rhythms (breeding, migration, flowering, and so on) in many organisms. Short-day plants flower when day-length is below a critical level. The cocklebur *(Xanthium strumarium),* a widespread weed in many parts of the world, flowers in spring when, as days become longer, a critical night-length is reached. Long-day plants flower when day-length is above a critical level. The strawberry tree *(Arbutus unedo)* flowers in the autumn as the night-length increases. In its Mediterranean home, this means that its flowers are ready for pollination when such long-tongued insects as bees are plentiful. Day-neutral plants flower after a period of vegetative growth, irrespective of the photoperiod.

In the high Arctic, plant growth is telescoped into a brief few months of warmth and light. Positive heliotropism (growing towards the Sun) is one way that plants can cope with limited light. It is common in Arctic and alpine flowers. The flowers of the Arctic avens *(Dryas integrifolia)* and the Arctic poppy *(Papaver radicatum)* track the Sun, turning at about 15° of arc per hour (Kevan 1975; see also Corbett *et al.* 1992). Their corollas reflect radiation onto their reproductive parts. The flowers of the alpine snow buttercup *(Ranunculus adoneus)* track the Sun's movement from early morning until mid-afternoon (Stanton and Galen 1989). Buttercup flowers aligned parallel to the Sun's rays reach mean internal temperatures several degrees Celsius above ambient air temperature. Internal flower temperature is significantly reduced as a flower's angle of deviation from the Sun increases beyond 45°.

Arctic and alpine animals and plants also have to cope with limited solar energy. Herbivores gear their behaviour to making the most of the short summer. Belding ground squirrels *(Spermophilus beldingi),* which live at high elevations in the western United States, are active for four or five summer months, and they must eat enough during that time to survive the winter on stored fat. To do this, their body temperature fluctuates by 3-4°C (to a high of 40°C) so that valuable energy is not wasted in keeping body temperature constant. Should they need to cool down, they go into a burrow or else adopt a posture that lessens exposure to sunlight. A constant breeze cools them during the hottest part of the day.

Some like it Hot : Temperature

Broadly speaking, average annual temperatures are highest at the equator and lowest at the poles. Temperatures also decrease with increasing elevation. The average annual temperature range is an important ecological factor. It is highest deep in high-latitude continental interiors and lowest over oceans, especially tropical oceans. In north-east Siberia, an annual temperature range of 60°C is not uncommon, whereas the range over equatorial oceans is less than about 3°C. Land lying adjacent to oceans, especially land on the western seaboard of continents, has an annual temperature range around the 11°C mark. These large differences in annual temperature range reflect differences in continentality (or oceanicity)—the winter temperatures of places near oceans will be less cold. Many aspects of temperature affect organisms, including daily, monthly, and annual extreme and mean temperatures, and the level of temperature variability. Different aspects of temperature are relevant to different species and commonly vary with the time of year and the stage in an organism's life cycle. Temperature may be limiting at any stage of an organism's life cycle. It may affect survival, reproduction, and the development of seedlings and young animals. It may affect competition with other organisms and susceptibility to predation, parasitism, and disease when the limits of temperature tolerance are approached. Many flowering plants are especially sensitive to low temperatures between germination and seedling growth.

Microbes and Temperature

Heat-loving microbes (thermophiles) reproduce or grow readily in temperatures over 45°C. Hyperthermophiles, such as *Sulfolobus acidocaldarius,* prefer temperatures above 80°C, and some thrive above 100°C. The most resistant hyperthermophile discovered to date is *Pyrolobus fumarii.* This microbe flourishes in the walls of 'smokers' in the deep-sea floor. It multiplies in temperatures up to 113°C. Below 94°C it finds it too cold and stops growing! Only in small areas that are intensely heated by volcanic activity do high temperatures prevent life. Cold-loving microbes (psychrophiles) are common in Antarctic sea ice. These communities include photosynthetic algae and diatoms, and a variety of bacteria. *Polaromonas vacuolata,* a bacterium, grows best at about 4°C, and stops reproducing above 12°C. Lichens can photosynthesize at –30°C, providing that they are not covered with snow. The reddish-coloured snow alga, *Chlamydomonas nivalis,* lives on ice and snow fields in the polar and nival zones, giving the landscape a pink tinge during the summer months.

Animals and Temperature

In most animals, temperature is a critical limiting factor. Vital metabolic processes are geared to work optimally within a narrow temperature band. Cold-blooded animals (poikilotherms) warm up and cool down with environmental temperature. They can assist the warming process a little by taking advantage of sunny spots or warm rocks. Most warm-blooded animals (homeotherms) maintain a constant body temperature amidst varying ambient conditions. They simply regulate the production and dissipation of heat. The terms 'cold-blooded' and 'warm-blooded' are misleading because the body temperature of some 'cold-blooded' animals may rise above that of 'warm-blooded' animals.

Each homeothermic species has a characteristic thermal neutral zone, a band of temperature within which little energy is expended in heat regulation. Small adjustments are made by fluffing or compressing fur, by making local changes in the blood supply, or by changing position. The bottom end of the thermal neutral zone is bounded by a lower critical temperature. Below this temperature threshold, the body's central heating system comes on fully. The colder it gets, the more oxygen is needed to burn fuel for heat. Animals living in cold environments are well insulated—fur and blubber can reduce the lower critical temperature considerably. An Arctic fox *(Alopex lagopus)* clothed in its winter fur rests comfortably at an ambient temperature of -50°C without increasing its resting rate of metabolism (Irving 1966). Below the lower critical temperature, the peripheral circulation shuts down to conserve energy. An Eskimo dog may have a deep body temperature of 38°C, the carpal area of the forelimb at 14°C, and foot pads at 0°C. Hollow hair is also useful for keeping warm. It is found in the American pronghorn *(Antilocapra americana),* an even-toed ungulate, and enables it to stay in open and windswept places at temperatures far below 0°C. The polar bear *(Ursus maritimus)* combines hollow hair, a layer of blubber up to 11 cm thick, and black skin to produce a superb insulating machine. Each hair acts like a fibre-optic cable, conducting warming ultraviolet light to the heat-absorbing black skin. This heating mechanism is so efficient that polar bears are more likely to overheat than to chill down, which partly explains their ponderousness. Many animals also have behavioural patterns designed to minimize heat loss. Some roll into a ball, some seek shelter. Herds of deer or elk seek ridge tops or south-facing slopes.

Above the upper critical temperature, animals must lose heat to prevent their overheating. Animals living in hot environments can lose much heat. Evaporation helps heat loss, but has an unwanted side-effect precious water is lost. Small animals can burrow to avoid high temperatures at the ground surface. In the Arizona desert, United States, most rodents burrow to a depth where hot or cold heat stress is not met with. Large size is an advantage in preventing overheating because the surface area is relatively greater than the body volume. Many desert mammals are adapted to high temperatures. The African rock hyrax *(Heterohyrax brucei)* has an upper critical temperature of 41°C. Camels (*Camelus* spp.), oryx (*Oryx* spp.), common eland *(Taurotragus oryx),* and gazelle (*Gazella* spp.) let their body temperatures fluctuate considerably over a 24-hour period, falling to about 35°C toward dawn and rising to over 40°C during the late afternoon. In an ambient temperature of 45°C sustained for 12 hours under experimental conditions, an oryx's temperature rose above 45°C and stayed there for 8 hours without injuring the animal. It has a specialized circulatory system that helps it to survive such excessive overheating.

Many mammal species are adapted to a limited range of environmental temperatures. Even closely related groups display significant differences in their ability to endure temperature extremes. The lethal ambient temperatures for four populations of woodrats (*Neotoma* spp.) in the western United States showed differences between species, and between populations of the same species living in different states.

Plants and Temperature

Temperature affects many processes in plants, including photosynthesis, respiration, growth, reproduction, and transpiration. Plants vary enormously in their ability to tolerate either heat or cold. There are five broad categories of cold tolerance. Chilling-sensitive plants, which are mostly tropical, are damaged by temperatures lower than 10°C. Chilling-resistant (frost-sensitive) plants can survive at temperatures below 10°C, but are damaged when ice forms within their tissues. Frost-resistant plants make physiological changes that enable them to survive temperatures as low as about “15°C. Frost-tolerant plants survive by withdrawing water from their cells, so preventing ice forming. The withdrawal of water also increases the concentration in sap and protoplasm, which acts as a kind of antifreeze, and lowers freezing point. Temperatures down to about “40°C can be tolerated in this way. Cold-tolerant plants, which are mostly needle-leaved, can survive almost any subzero temperature.

Cold tolerance varies enormously at different seasons in some species. Willow twigs *(Salix* spp.) collected in winter can survive freezing temperatures below “150°C; the same twigs in summer are killed by a temperature of “5°C. Similarly, the red-osier dogwood *(Cornus stolonifera),* a hardy shrub from North America, could survive a laboratory test at “196°C by midwinter when grown in Minnesota (Weiser 1970). Nonetheless, dogwoods native to coastal regions with mild climates are often damaged by early autumn frosts. Plants growing on Mt Kurodake, Hokkaido Province, Japan, are killed by temperatures of “5°C to “7°C during the growing season. In winter, most of the same plants survive freezing to “30°C, and the willow ezo-mame-yanagi *(Salix pauciflora),* mosses, and lichens will withstand a temperature of “70°C. Acclimatization or cold hardening accounts for these differences. The coastal dogwoods did not acclimatize quickly enough. Timing is important in cold resistance, but absolute resistance can be altered. Many plants use the signal of short days in autumn as an early warning system. The short days trigger metabolic changes that stop the plant growing and produce resistance to cold. Many plant species, especially deciduous plants in temperate regions, need chilling during winter if they are to grow well the following summer. Chilling requirements are specific to species. They are often necessary for buds to break out of dormancy, a process called *vernalization.*

Many plants require a certain amount of 'warmth' during the year. The total 'warmth' depends on the growing season length and the growing season temperature. These two factors are combined as 'day-degree totals'. Day-degree totals are the product of the growing season length (the number of days for which the mean temperature is above standard temperature, such as freezing point or 5°C), and the mean temperature for that period. The Iceland purslane *(Koenigia islandica),* a tundra annual, needs only 700 day-degrees to develop from a germinating seed to a mature plant producing seeds of its own. The small-leaved lime *(Tilia cordata),* a deciduous tree, needs 2,000 day-degrees to complete its reproductive development. Trees in tropical forests may need up to 10,000 day-degrees to complete their reproductive development.

Excessive heat is as detrimental to plants as excessive cold. Plants have evolved resistance to heat stress, though the changes are not so marked as resistance to cold stress. Different parts of plants acquire differing degrees of heat resistance, but the pattern varies between species. In some species the uppermost canopy leaves are often the most heat resistant; in other species, it is the middle canopy leaves, or the leaves at the base of the plant. Temperatures of about 44°C are usually injurious to evergreens and shrubs from cold-winter regions. Temperate-zone trees are damaged at 50-55°C, tropical trees at 45-55°C. Damage incurred below about 50°C can normally be repaired by the plant; damage incurred above that temperature is most often irreversible. Exposure time to excessive heat is a critical factor in plant survival, while exposure time to freezing temperatures is not.

Quenching Thirst: Moisture

Protoplasm, the living matter of animal and plant cells, is about 90 per cent water—without adequate moisture there can be no life. Water affects land animals and plants in many ways. Air humidity is important in controlling loss of water through the skin, lungs, and leaves. All animals need some form of water in their food or as drink to run their excretory systems. Vascular plants have an internal plumbing system—parallel tubes of dead tissue called xylem—that transfers water from root tips to leaves. The entire system is full of water under stress (capillary pressure). If the water stress should fall too low, disaster may ensue—germination may fail, seedlings may not become established, and, should the fall occur during flowering, seed yields may be severely cut. An overlong drop in water stress kills plants, as anybody who has tried to grow bedding plants during a drought and hose-pipe ban will know.

Bioclimates

On land, precipitation supplies water to ecosystems. Plants cannot use all the precipitation that falls. A substantial portion of the precipitation evaporates and returns to the atmosphere. For this reason, available moisture (roughly the precipitation less the evaporation) is a better guide than precipitation to the usable water in a terrestrial ecosystem. This point is readily understood with an example. A mean annual rainfall of 400 mm might support a forest in Canada, where evaporation is low, but might support a dry savannah in Tanzania, where evaporation is high.

Available moisture largely determines soil water levels, which in turn greatly influence plant growth. For a plant to use energy for growth, water must be available. Without water, the energy will merely heat and stress the plant. Similarly, for a plant to use water for growth, energy must be obtainable. Without an energy source, the water will run into the soil or run off unused.

For these reasons, temperature (as a measure of energy) and moisture are master limiting factors that act in tandem. In tropical areas, temperatures are always high enough for plant growth and precipitation is the limiting factor. In cold environments, water is usually available for plant growth for most of the year—low temperatures are the limiting factor. This is true, too, of limiting factors on mountains where lower altitudinal limits are set by heat or water or both, and upper altitudinal limits are set by a lack of heat.

So important are precipitation and temperate that several researchers use them to characterize bioclimates. Bioclimates are the aspects of climate that seem most significant to living things. The most widely used bioclimatic classification is the 'climate diagram' devised by Heinrich Walter. This is the system of summarizing ecophysiological conditions that makes David Bellamy 'feel like a plant'.

Climate diagrams portray climate as a whole, including the seasonal round of precipitation and temperature. They show at a glance the annual pattern of rainfall; the wet and dry seasons characteristic of an area, as well as their intensity, since the evaporation rate is directly related to temperature; the occurrence or non-occurrence of a cold season, and the months in which early and late frost have been recorded. Additionally, they provide information on such factors as mean annual temperature, mean annual precipitation, the mean daily minimum temperature during the coldest month, the absolute minimum recorded temperature, the altitude of the station, and the number of years of record.

Wet Environments

Plants are very sensitive to water levels. Hydrophytes are water plants and root in standing water. Helophytes are marsh plants. Mesophytes are plants that live in normally moist but not wet conditions. Xerophytes are plants that live in dry conditions. Wetlands support hydrophytes and helophytes. The common water crowfoot *(Ranunculus aquatalis)* and the bog pondweed *(Potamogeton polygonifolius)* are hydrophytes; the greater bird's-foot trefoil *(Lotus uliginosus)* is a helophyte. These plants manage to survive by developing a system of air spaces in their roots, stems, or leaves. The air spaces provide buoyancy and improve internal ventilation. Mesophytes vary greatly in their ability to tolerate flooding. In the southern United States, bottomland hardwood forests occupy swamps and river floodplains. They contain a set of tree species that can survive in a flooded habitat. The water tupelo *(Nyssa aquatica),* which is found in bottomland forest in the southeastern United States, is well adapted to such wet conditions.

PLANT LIFE FORMS

Phanerophytes

Phanerophytes (from the Greek *phaneros,* meaning visible) are trees and large shrubs. They bear their buds on shoots that project into the air and are destined to last many years. The buds are exposed to the extremes of climate. The primary shoots, and in many cases the lateral shoots as well, are negatively geotropic (they stick up into the air). Weeping trees are an exception. Raunkiaer divided phanerophytes into twelve subtypes according to their bud covering (with bud-covering or without it), habit (deciduous or evergreen), and size (mega, meso, micro, and nano); and three other subtypes herbaceous phanerophytes, epiphytes, and stem succulents. A herbaceous example is the scaevola, *Scaevola koenigii.* Phanerophytes are divided into four size classes: megaphanerophytes (> 30 m), mesophanerophytes (8-30 m), microphanerophytes (2-8 m) and nanophanerophytes (< 2 m).

Chomaephytes

Chamaephytes (from the Greek *khamai,* meaning on the ground) are small shrubs, creeping woody plants, and herbs. They bud from shoot-apices very close to the ground. The flowering shoots project freely into the air but live only during the favourable season. The persistent shoots bearing buds lie along the soil, rising no more than 20-30 cm above it.

Suffructicose chamaephytes have erect aerial shoots that die back to the ground when the unfavourable season starts. They include species of the Labiatae, Caryophyllaceae, and Leguminosae. Passive chamaephytes have procumbent persistent shoots—they are long, slender, comparatively flaccid, and heavy and so lie along the ground. Examples are the greater stitchwort *(Stellaria holostea)* and the prostrate speedwell (*Veronica prostrata*). Active chamaephytes have procumbent persistent shoots that lie along the ground because they are transversely geotropic in light (take up a horizontal position in response to gravity). Examples are the heath speedwell *(Veronica officinalis)*, the crowberry *(Empetrum nigrum)*, and the twinflower *(Linnaea borealis)*. Cushion plants are transitional to hemicryptophytes. They have very low shoots, very closely packed together. Examples are the hairy rock-cress *(Arabis hirusa)* and the houseleek *(Sempervivum tectorum)*.

Hemicryptophytes

Hemicryptophytes (from the Greek *kryptos,* meaning hidden) are herbs growing rosettes or tussocks. They bud from shoot-apices located in the soil surface. They include protohemicryptophytes (from the base upwards, the aerial shoots have elongated internodes and bear foliage leaves) such as the vervain *(Verbena officinalis)*, partial rosette plants such as the bugle *(Ajuga reptans)*, and rosette plants such as the daisy *(Bellis perennis)*.

Cryptophytes

Cryptophytes are tuberous and bulbous herbs. They are even more 'hidden' than hemicryptophytes—their buds are completely buried beneath the soil, thus affording them extra protection from freezing and drying. They include geophytes (with rhizomes, bulbs, stem tubers, and root tuber varieties) such as the purple crocus *(Crocus vernus)*, helophytes or marsh plants such as the arrowhead *(Sagittaria sagittifolia)*, and hydrophytes or water plants such as the rooted shining pondweed *(Potamogeton lucens)* and the free-swimming frogbit *(Hydrocharis morsus-ranae)*.

Therophytes

Therophytes (from the Greek *theros,* meaning summer) or annuals are plants of the summer or favourable season and survive the adverse season as seeds. Examples are the cleavers *(Galium aparine)*, the cornflower *(Centaurea cyanus)*, and the wall hawk's-beard *(Crepis tectorum)*.

An example of the 'autoecological accounts' for the bluebell *(Hyacinthoides non-scripta*. The bluebell is a polycarpic perennial, rosette-forming geophyte, with a deeply buried bulb. It appears above ground in the spring, when it exploits the light phase before the development of a full summer canopy. It is restricted to sites where the light intensity does not fall below 10 per cent of the daylight between April and mid-June, in which period the flowers are produced. Shoots expand during the late winter and early spring. The seeds are gradually shed, mainly in July and August. The leaves are normally dead by July. There is then a period of aestivation (dormancy during the dry season). This ends in the autumn when a new set of roots forms. The plant cannot replace damaged leaves and is very vulnerable to grazing, cutting, or trampling. Its foliage contains toxic glycosides and, though sheep and cattle will eat it, rabbits will not. Its reproductive strategy is intermediate between a stress-tolerant ruderal and a competitor—stress-tolerator—ruderal 103). It extends to 340 m around Sheffield, but is known to grow up to 660 m in the British Isles. It is largely absent from skeletal habitats and steep slopes. The bluebell commonly occurs in woodland. In the Sheffield survey, it was recorded most frequently in broad-leaved plantations. It was also common in scrub and woodland overlying either acidic or limestone beds, but less frequent in coniferous plantations. It occurs in upland areas on waste ground and heaths, and occasionally in unproductive pastures, on spoil heaps, and on cliffs. In woodland habitats, it grows more frequently and is significantly more abundant on south-facing slopes. However, in unshaded habitats, it prefers north-facing slopes. It is not found in wetlands. It can grow on a wide range of soils, but it most frequent and more abundant in the pH range 3.5-7.5. It is most frequent and abundant in habitats with much tree litter and little exposed soil, though it is widely distributed across all bare-soil classes.

ADAPTING TO CIRCUMSTANCES: NICHES AND LIFE-FORMS

Ways of Living

Organisms have evolved to survive in the varied conditions found at the Earth's surface. They have come to occupy nearly all habitats and to have filled multifarious roles within food chains.

Ecological Niche

An organism's ecological niche (or simply niche) is its 'address' and 'profession'. Its address or home is the habitat in which it lives, and is sometimes called the habitat niche. Its profession or occupation is its position in a food chain, and is sometimes called the functional niche. A skylark's *(Alauda arvensis)* address is open moorland (and, recently, arable farmland); its profession is insect-cum-seed-eater. A merlin's *(Falco columbarius)* address is open country, especially moorland; its profession is a bird-eater, with skylark and meadow pipit *(Anthus pratensis)* being its main prey. A grey squirrel's *(Sciurus carolinensis)* habitat niche is a deciduous woodland; its profession is a nut-eater (small herbivore). A grey wolf's *(Canis lupus)* habitat niche is cool temperate coniferous forest, and its profession is large-mammal-eater.

A distinction is drawn between the fundamental niche and the realized niche. The fundamental (or virtual) niche circumscribes where an organism would live under optimal physical conditions and with no competitors or predators. The realized (or actual) niche is always smaller, and defines the 'real-world' niche occupied by an organism constrained by biotic and abiotic limiting factors.

A niche reflects how an individual, species, or population interacts with and exploits its environment. It involves adaptation to environmental conditions. The competitive exclusion principle precludes two species occupying identical niches. However, a group of species, or guild, may exploit the same class of environmental resources in a similar way. In an oak woodland, one guild of birds forages for arthropods from the foliage of oak trees; another catches insects in the air; another eats seeds. The foliage-gleaning guild in a California oak woodland includes members of four families: the plain titmouse (*Parus inornatus,* Paridae), the blue-gray gnatcatcher (*Polioptila caerulea,* Sylviidae), the warbling vireo and Hutton's vireo (*Vireo gilvus* and *Vireo huttoni,* Vireonidae), and the orange-crowned warbler (*Vermivora celata,* Parulidae).

Ecological Equivalents

Although each niche is occupied by only one species, different species may occupy the same or similar niches in different geographical regions. These species are ecological equivalents or vicars. A grassland ecosystem contains a niche for large herbivores living in herds. Bison and the pronghorn antelope occupy this niche in North America; antelopes, gazelles, zebra, and eland in Africa; wild horses and asses in Europe; the pampas deer and guanaco in South America; and kangaroos and wallabies in Australia. As this example shows, quite distinct species may become ecological equivalents through historical and geographical accidents.

Many bird guilds have ecological equivalents on different continents. The nectar-eating (nectivore) guild has representatives in North America, South America, and Africa. In Chile and California the representatives are the hummingbirds (Trochilidae) and the African representatives are the sunbirds (Nectariniidae). One remarkable convergent feature between hummingbirds and sunbirds is the iridescent plumage.

Plant species of very different stock growing in different areas, when subjected to the same environmental pressures, have evolved the same life-form to fill the same ecological niche. The American cactus and the South African euphorbia, both living in arid regions, have adapted by evolving fleshy, succulent stems and by evolving spines instead of leaves to conserve precious moisture.

Life-forms

The structure and physiology of plants and, to a lesser extent, animals are often adapted for life in a particular habitat. These structural and physiological adaptations are reflected in life-form and often connected with particular ecozones. The life-form of an organism is its shape or appearance, its structure, its habits, and its kind of life history. It includes overall form (such as herb, shrub, or tree in the case of plants) and the form of individual features (such as leaves). Importantly, the Plant life-forms.

A widely used classification of plant life-forms, based on the position of the shoot-apices (the tips of branches) where new buds appear, was designed by Christen Raunkiaer in 1903. It distinguishes five main groups: therophytes, cryptophytes, hemicryptophytes, chamaephytes, and phanerophytes.

A biological spectrum is the percentages of the different life-forms in a given region. The 'normal spectrum' is a kind of reference point; it is the percentages of different life-forms in the world flora. Each ecozone possesses a characteristic biological spectrum that differs from the 'normal spectrum'. Tropical forests contain a wide spectrum of life-forms, whereas in extreme climates, with either cold or dry seasons, the spectrum is smaller. As a rule of thumb, very predictable, stable climates, such as humid tropical climates, support a wider variety of plant life-forms than do regions with inconstant climates, such as arid, Mediterranean, and alpine climates. Alpine regions, for instance, lack trees, the dominant life-form being dwarf shrubs (chamaephytes). In the Grampian Mountains, Scotland, 27 per cent of the species are chamaephytes. Some life-forms appear to be constrained by climatic factors. Megaphanerophytes (where the regenerating parts stand over 30 m from the ground) are found only where the mean annual temperature of the warmest month is 10°C or more. Trees are confined to places where the mean summer temperature exceeds 10°C, both altitudinally and latitudinally. This uniform behaviour is somewhat surprising as different taxa are involved in different countries. Intriguingly, dwarf shrubs, whose life cycles are very similar to those of trees, always extend to higher altitudes and latitudes than do trees.

Individual parts of plants also display remarkable adaptations to life in different ecozones. This is very true of leaves. In humid tropical lowlands, forest trees have evergreen leaves with no lobes. In regions of Mediterranean climate, plants have small, sclerophyllous evergreen leaves. In arid regions, stem succulents without leaves, such as cacti, and plants with entire leaf margins have evolved. In cold wet climates, plants commonly possess notched or lobed leaf margins.

Animal Life-forms

Animal life-forms, unlike those of plants, tend to match taxonomic categories rather than ecozones. Most mammals are adapted to basic habitats and may be classified accordingly. They may be adapted for life in water (aquatic or swimming mammals), underground (fossorial or burrowing mammals), on the ground (cursorial or running, and saltatorial or leaping mammals), in trees (arboreal or climbing mammals), and in the air.

Autoecological Accounts

Detailed habitat requirements of individual species require careful and intensive study. A ground-breaking study was the autoecological accounts prepared for plants around Sheffield, England. About 3,000 km^2 were studied in three separate surveys by the Natural Environment Research Council's Unit of Comparative Plant Ecology (formerly the Nature Conservancy Grassland Research Unit). The region comprises two roughly equal portions: an 'upland' region, mainly above 200 m and with mean annual precipitation more than 850 mm, underlain by Carboniferous Limestone, Millstone Grit, and Lower Coal Measures; and a drier, 'lowland' region overlying Magnesian Limestone, Bunter Sandstone, and Keuper Marl.

SOILS, SLOPES, AND DISTURBING AGENCIES

Soil and substrate influence animals and plants, both at the level of individual species and at the level of communities.

Microbes and Substrate

Acid-loving microbes (acidophiles) prosper in environments with a pH below 5. *Sulfolobus acidocaldarius,* as well as liking it hot, also likes it acid. Alkali-loving microbes (alkaliphiles) prefer an environment with a pH above 9. *Natronobacterium gregoryi* lives in soda lakes. Salt-loving microbes live in intensely saline environments. They survive by producing large amounts of internal solutes that prevents rapid dehydration in a salty medium. An example is *Halobacterium salinarium*.

Plants and Substrate

Plants seem capable of adapting to the harshest of substrates. Saxicolous vegetation grows on cliffs, rocks, and screes, some species preferring rock crevices (chasmophytes), others favouring small ledges where detritus and humus have collected (chomophytes). In the Peak District of Derbyshire, England, maidenhair spleenwort *(Asplenium trichomanes)* is a common chasmophyte and the wallflower *(Cheiranthus cheiri)* is a common and colourful chomophyte. Perhaps the most extreme adaptation to a harsh environment is seen in the mesquite trees (*Prosopis tamarugo* and *Prosopis alba*) that grow in the Pampa del Tamagural, a closed basin, or salar, in the rainless region of the Atacama Desert, Chile. These plants manage to survive on concrete-like carbonate surfaces. Their leaves abscise (are shed) and accumulate to depths of 45 cm. Because there is virtually no surface water, the leaves do not decompose and nitrogen is not incorporated back into the soil for recycling by plants. The thick, crystalline pan of carbonate salts prevents roots from growing into the litter. To survive, the trees have roots that fix nitrogen in moist subsurface layers, and extract moisture and nutrients from groundwater at depths of 6-8 m or more through a tap root and a mesh of fine roots lying between 50 and 200 cm below the salt crust. A unique feature of this ecosystem is the lack of nitrogen cycling.

Calcicoles (or calciphiles) are plants that favour such calcium-rich rocks as chalk and limestone. Calcicolous species often grow only on soil formed in chalk or limestone. An example from England, Wales, and Scotland is the meadow oat-grass *(Helictotrichon pratense)*, the distribution of which picks out the areas of chalk and limestone and the calcium-rich schists of the Scottish Highlands. Other examples are traveller's joy *(Clematis vitalba)*, the spindle tree *(Euonymus europaeus)*, and the common rock-rose *(Helianthemum nummularium)*. Calcifuges (or calciphobes) avoid calcium-rich soils, preferring instead acidic soils developed on rocks deficient in calcium. An example is the wavy hair-grass *(Deschampsia flexuosa)*. However, many calcifuges are seldom entirely restricted to exposures of acidic rocks. In the limestone Pennine dales, the wavy hair-grass can be found growing alongside meadow oat-grass. Neutrophiles are acidity 'middle-of-the-roaders'. They tend to grow in the range pH 5-7. In the Pennine dales, strongly growing, highly competitive grasses that make heavy demands on water and nutrient stores are the most common neutrophiles.

Animals and Substrate

Some animals are affected by soil and substrate. For instance, the type and texture of soil or substrate is critical to two kinds of mammals: those that seek diurnal refuge in burrows, and those that have modes of locomotion suited to relatively rough surfaces. Burrowing species, which tend to be small, may be confined to a particular kind of soil. For instance, many desert rodents display marked preferences for certain substrates. In most deserts, no single species of rodent is found on all substrates; and some species occupy only one substrate. Four species of pocket mice *(Perognathus)* live in Nevada, United States. Their preferences for soil types are largely complementary: one lives on fairly firm soils of slightly sloping valley margins; the second is restricted to slopes where stones and cobbles are scattered and partly embedded in the ground; the third is associated with the fine, silty soil of the bottomland; and the fourth, a substrate generalist, can survive on a variety of soil types.

Saxicolous species grow in, or live among, rocks. Some woodrats *(Neotoma)* build their homes exclusively in cliffs or steep rocky outcrops. The dwarf shrew *(Sorex nanus)* seems confined to rocky areas in alpine and subalpine environments. Even some saltatorial species are adapted to life on rocks. The Australian rock wallabies (*Petrogale* and *Petrodorcas*) leap adroitly among rocks. They are aided in this by traction-increasing granular patterns on the soles of their hind feet. Rocky Mountain pikas *(Ochotona princeps)* in the southern Rocky Mountains, United States, normally live on talus or extensive piles of gravel. Those living near Bodie, a ghost town in the Sierra Nevada, utilize tailings of abandoned gold mines. The yellow-bellied marmot *(Marmota flaviventris)* is another saxicolous species, and commonly occurs with the Rocky Mountain pika. The entire life style of African rock hyraxes *(Heterohyrax, Procavia)* is built around their occupancy of rock piles and cliffs. Most of their food consists of plants growing among, or very close by, rocks. Their social system is bonded by the scent of urine and faeces on the rocks. The rocks provide useful vantage points to keep an eye out for predators, hiding places, and an economical means of conserving energy.

Pedobiomes

Within zonobiomes, there are areas of intrazonal and azonal soils that, in some cases, support a distinctive vegetation. These non-zonal vegetation communities are pedobiomes. Several different pedobiomes are distinguished on the basis of soil type: lithobiomes on stony soil, psammobiomes on sandy soil, halobiomes on salty soil, helobiomes in marshes, hydrobiomes on waterlogged soil, peinobiomes on nutrient-poor soils, and amphibiomes on soils that are flooded only part of the time. Pedobiomes commonly form a mosaic of small areas and are found in all zonobiomes.

There are instances where pedobiomes are extensive: the Sudd marshes on the White Nile, which cover 150,000 km^2; fluvioglacial sandy plains; and the nutrient poor soils of the Campos Cerrados in Brazil.

A striking example of a lithobiome is found on serpentine. The rock serpentine and its relatives, the serpentinites, are deficient in aluminium. This leads to slow rates of clay formation, which explains the characteristic features of soils formed on serpentinites: they are high erodible, shallow, and stock few nutrients. These peculiar features have an eye-catching influence on vegetation. Outcrops of serpentine support small islands of brush and bare ground in a sea of forest and grassland. These islands are populated by native floras with many endemic species.

TERRESTRIAL ECOZONES

On land, characteristic animal and plant communities are associated with nine basic climatic types, variously called zonobiomes, ecozones, and ecoregions.

Polar and Subpolar Zone

This zone includes the Arctic and Antarctic regions. It is associated with tundra vegetation. The Arctic tundra regions have low rainfall evenly distributed throughout the year. Summers are short, wet, and cool. Winters are long and cold. Antarctica is an icy desert, although summer warming around the fringes is causing it to bloom.

Boreal Zone

This is the cold-temperate belt that supports coniferous forest (taiga). It usually has cool, wet summers and very cold winters lasting at least six months. It is found only in the Northern Hemisphere where it forms a broad swath around the pole—it is a circumpolar zone.

Humid Mid-latitude Zone

This zone is the temperate or nemoral zone. In continental interiors it has a short, cold winter and a warm, or even hot, summer. Oceanic regions, such as the British Isles, have warmer winters and cooler, wetter summers. This zone supports broad-leaved deciduous forests.

Arid Mid-latitude Zone

This is the cold-temperate (continental) belt. The difference between summer and winter temperatures is marked and rainfall is low. Regions with a dry summer but only a slight drought support temperate grasslands. Regions with a clearly defined drought period and a short wet season support cold desert and semi-desert vegetation.

Tropical and Subtropical Arid Zone

This is a hot desert climate that supports thorn and scrub savannahs and hot deserts and semi-deserts.

Mediterranean Subtropical Zone

This is a belt lying between roughly 35° and 45° latitude in both hemispheres with winter rains and summer drought. It supports sclerophyllous (thick-leaved), woody vegetation adapted to drought and sensitive to prolonged frost.

Seasonal Tropical Zone

This zone extends from roughly 25° to 30° North and South. There is a marked seasonal temperature difference. Heavy rain in the warmer summer period alternates with extreme drought in the cooler winter period.

The annual rainfall and the drought period increase with distance from the equator. The vegetation is tropical grassland or savannah.

Humid Subtropical Zone

This zone has almost no cold winter season, and short wet summers. It is the warm temperate climate in Walter's zonobiome classification. Vegetation is subtropical broad-leaved evergreen forest.

Humid Tropical Zone

This torrid zone has rain all year and supports evergreen tropical rain forest. The climate is said to be diurnal because it varies more by day and night than it does through the seasons.

Marine ecozones

The marine biosphere also consists of 'climatic' zones, which are also called ecozones. The main surface-water marine ecozones are the polar zone, the temperate zone, and the tropical zone:

1. *Polar Zone*: Ice covers the polar seas in winter. Polar seas are greenish, cold, and have a low salinity.
2. *Temperate Zone*: Temperate seas are very mixed in character. They include regions of high salinity in the subtropics.
3. *Tropical Zone*: Tropical seas are generally blue, warm, and have a high salinity.

Biomes

Each ecozone supports several characteristic communities of animals and plants known as biomes. The deciduous forest biome in temperate western Europe is an example. It consists largely of woodland with areas of heath and moorland. A plant community at the biome scale—all the plants associated with the deciduous woodland biome, for example—is a plant formation. A n equivalent animal community has no special name; it is simply an animal community. Smaller communities within biomes are usually based on plant distribution. They are called plant associations. In England, associations within the deciduous forest biome include beech forest, lowland oak forest, and ash forest. Between biomes are transitional belts where the climate changes from one type to the next. These are called ecotones.

Zonobiomes

All the biomes around the world found in a particular ecozone constitute a zonobiome. A plant community at the same large scale is a formation-type or zonal plant formation. The broad-leaved temperate forests of western Europe, North America, eastern Asia, southern Chile, south-east Australia and Tasmania, and most of New Zealand comprise the humid temperate zonobiome. Between the zonobiomes are transitional belts where the climate changes from one type to the next. These are called zonoecotones.

Freshwater communities (lakes, rivers, marshes, and swamps) are part of continental zonobiomes. They may be subdivided in various ways. Lakes, for instance, may be well mixed (polymictic or oligomictic) or permanently layered (meromictic). They may be wanting in nutrients and biota (oligotrophic) or rich in nutrients and algae (eutrophic). A thermocline (where the temperature profile changes most rapidly) separates a surface-water layer mixed by wind (epilimnon) from a more sluggish, deep-water layer (hypolimnon). And, as depositional environments, lakes are divided into a littoral (near-shore) zone, and a profundal (basinal) zone.

Marine ecozones, and the deep-water regions, consist of biomes (equivalent to terrestrial zonobiomes). The chief marine biomes are the intertidal (estuarine, littoral marine, algal bed, coral reef) biome, the open sea (pelagic) biome, the upwelling zone biome, the benthic biome, and the hydrothermal vent biome.

Orobiomes

Mountain areas possess their own biomes called orobiomes. The basic environmental zones seen on ascending a mountain are submontane (colline, lowland), montane, subalpine, alpine, and nival. On south-facing slopes in the Swiss Alps around Cortina, the submontane belt lies below about 1,000 m. It consists of oak forests and fields. The montane belt ranges from about 1,000 m to 3,000 m. The bulk of it is Norway spruce *(Picea abies)* forest, with scattered beech *(Fagus sylvatica)* trees at lower elevations. Mountain pines *(Pinus montana)* with scattered Swiss stone pines *(Pinus cembra)* grow near the tree line. The subalpine belt lies between about 3,000 m and 3,500 m. It contains diminutive forests of tiny willow (*Salix* spp.) trees, only a few centimetres tall when mature, within an alpine grassland. The alpine belt, which extends up to about 4,000 m, is a meadow of patchy grass and a profusion of alpine flowers—poppies, gentians, saxifrages, and many more. The mountain tops above about 4,000 m lie within the nival zone and are covered with permanent snow and ice.

CLASSIFYING ORGANISMS

Everyone knows that living things come in a glorious diversity of shapes and sizes. It is apparent, even to a casual observer, that organisms appear to fall into groups according to the similarities between them. No one is likely to mistake a bird for a beetle, a daisy for a hippopotamus. Zoologists and botanists classify organisms according to the similarities and differences between them. Currently, five great kingdoms are recognized prokaryotae (monera), protoctista, plantae, fungi, and animalia. These chief subdivisions of the kingdoms are phyla. Each phylum represents a basic body plan that is quite distinct from other body plans. This is why it is fairly easy, with a little practice, to identify the phylum to which an unidentified organism belongs.

Organisms are classified hierarchically. Individuals are grouped into species, species into genera, genera into families, and so forth. Each species, genus, family, and higher-order formal group of organisms is called a taxon (plural taxa). Each level in the hierarchy is a taxonomic category. The following list shows the classification of the ring ouzel:

Animal family names always end in -idae, and sub-families in -inae; they may be less formally referred to by dropping the initial capital letter and using -ids as an ending, as in felids for members of the cat family. Plant family names end in -aceae or -ae. The genus (plural genera) is the first term of a binomial: genus plus species, as in *Turdus torquatus*. It is always capitalized and in italics.

The species is the second term of a binomial. It is not capitalized in animal species, and is not normally capitalized in plant species, but is always italicized in both cases. The specific name signifies either the person who first described it, as in *Muntiacus reevesi,* Reeve's muntjac deer, or else some distinguishing feature of the species, as in *Calluna vulgaris,* the common (=vulgar) heather. If subspecies are recognized, they are denoted by the third term of a trinomial. For example, the common jay in western Europe is *Garrulus glandarius glandarius,* which would usually be shortened to *Garrulus g. glandarius*. The Japanese subspecies is *Garrulus glandarius japonicus*. In formal scientific writing, the author or authority of the name is indicated. So, the badger's full scientific name is *Meles meles* L., the L. indicating that the species was first described by Carolus Linnaeus.

After its first appearance in a paper or book, the species name is usually abbreviated by reducing the generic term to a single capital letter. Thus, *Meles meles* becomes *M. meles*. This practice will not be adopted in the present book because it gives a rather 'stuffy' feel to the text. Likewise, the authorities will be omitted because they confer an even stuffier feel.

Nestor Parrots

The nestor parrots (Nestorinae) are endemic to New Zealand. There are two species—the kaka (*Nestor meridionalis)* and the kea *(Nestor notabilis*. They are closely related and are probably descended from a 'proto-kaka' that reached New Zealand during the Tertiary period. Then, New Zealand was a single, forest-covered island. The proto-kaka became adapted to forest life. Late in the Tertiary period, the north and south parts of New Zealand split. North Island remained forested and the proto-kakas there continued to survive as forest parrots, feeding exclusively on vegetable matter and nesting in tree hollows. They eventually evolved into the modern kakas. South Island gradually lost its forests because mountains grew and climate changed. The proto-kakas living on South Island adjusted to these changes by becoming 'mountain parrots', depending on alpine shrubs, insects, and even carrion for food. They forsook trees as breeding sites and turned to rock fissures. The changes in the South Island proto-kakas were so far-reaching that they became a new species—the kea. After the Ice Age, climatic amelio-ration promoted some reforestation of South Island. The kakas dispersed across the Cook Strait and colonized South Island. Interaction between North and South Island kaka populations is difficult across the 26 km of ocean. In consequence, the South Island kakas have become a subspecies. The kaka and the kea are now incapable of interbreeding and they continue to live side by side on South Island. The kea has never colonized North Island, probably because there is little suitable habitat there. The biogeography of the nestor parrots thus involves adaptation to changing environmental conditions, dispersal, and vicariance events.

Marsupials

The pouched mammals or marsupials are now found in Australia, South America, and North America. Fossil forms are known from Eurasia, North Africa, and Antarctica. They are commonly assumed to have evolved in North America, where the oldest known marsupial fossils are found, from an ancestor that also sired the placental mammals. The marsupial and placental mammals split about 130 million years ago, early in the Cretaceous period. There are several rival explanations for the current distribution of marsupials (L.G. Marshall 1980).

The classic explanation of marsupial distribution was proposed before continental drift was accepted and assumed stationary continents. It argued that marsupials dispersed from a Cretaceous North American homeland to other continents. Some time in the Late Cretaceous period, marsupials hopped across islands linking North and South America. During the Eocene epoch, they moved into Asia and Europe across a land bridge spanning the Bering Sea between Alaska and Siberia: From there they spread into Europe and, using Indonesia as an embarkation point, into Australia. Several variations on this 'centre of origin followed by dispersal' hypothesis played out on a stationary land surface were forthcoming. The variations involved different centres of origin (South America or Antarctica) and different dispersal routes.

As soon as it was accepted that the continents do drift, revised explanations of the marsupial history were suggested. Some of these new hypotheses still invoked centre-of-origin and dispersal. They were similar to the hypothesis developed for stationary continents, but they did not need to invoke fanciful land bridges between widely separated continents. Other hypotheses laid emphasis on the fragmentation of Pangaea, the Triassic supercontinent, and stressed vicariance events rather than dispersal over pre-existing barriers. It now seems likely that, even if the first marsupials did appear in Mesozoic North America (and that is far from certain), they quickly became widely distributed over the connected land masses of South America, Antarctica, and Australia. The breakup of Pangaea, which started in earnest during the Mid Jurassic period, isolated the Mesozoic marsupials on South America and Australia and these two main branches then evolved independently. The American marsupials reached Europe, North Africa, and Asia in Palaeocene and Eocene times, but by the end of the Miocene epoch, North American and European marsupials were extinct. South American marsupials invaded North America in the Pleistocene epoch. Current explanations of marsupial biogeography thus call on dispersal and vicariance events.

Biogeographical Regions

By the nineteenth century, it was abundantly clear that the land surface could be divided into several large biogeographical regions, each of which supports a distinct set of animals and a distinct set of plants. The Old World he divided into Europe and northern Asia, Africa south of the Sahara, India and southern Asia, and Australia and New Guinea. The New World he divided into North America and South America. Sclater's system was adopted by Alfred Russel Wallace, with minor amendments, to provide a long-lasting nomenclature that survives today as the Sclater Wallace scheme. Six regions are recognized Nearctic, Neotropical, Palaearctic, Ethiopian, Oriental, and Australian. Together, the Nearctic and Palaearctic form Neogaea (the New World), while other regions form Palaeogaea (The Old World). Wallace also recognized subregions, four per region which correspond largely to established plant regions. Modern methods of numerical classification produce similar regions to the Sclater Wallace scheme, but there are differences. The faunal regions of the world based on mammal distributions. There are four regions Holarctic, Latin American, Afro-Tethyan, and Island and ten subregions. Each subregion is as unique as it can be compared with all other subregions.

2

Biogeography: Global Patterns and Timing of Diversification

Our results provide another example of extensive cryptic diversity in species with Indo-Pacific distribution ranges. Nerita species from IAA present numerous robust divergent clades within this region. This pattern has been interpreted as a consequence of intense diversification within this region, producing species that further disperse and colonize peripheral islands in the Indian and Pacific oceans. Long distance colonization events may lead to allopatric and peripatric speciation with the modification of gene flow through time due to changes in biotic and abiotic conditions. Allopatry seems to be the most frequent speciation mechanism of marine species, but other processes like disruptive selection, habitat or resource choice, may occur at smaller geographic scales and lead to sympatric sister species, particularly in gastropods.

The first fossil record of Nerita is from the late Paleocene (56 Ma) and the average genus diversification rate varied over time according to the TreePar analysis. Our results suggest that diversification was higher during the Oligocene and early Miocene compared to later geologic periods when net diversification rate decreased from 0.058 to 0.00013. During Oligo-Miocene, the oceanic circulation of the southern hemisphere was greatly modified by the northward movement of the Australian and South-American plates and the southward migration of the Antarctic plate. These led to the formation of the circumpolar current, the formation of the Antarctic ice sheet and global cooling, lowering sea levels and expanding emerged land masses and coastlines.

Furthermore, the collision of the Australian and Asian plates led to the emergence of new landmasses and tropical habitats suitable for colonization by shallow water species: the IAA. The IAA formation modified equatorial currents, constraining seaways between the Pacific and Indian oceans. These changes in marine habitats induced massive species extinctions in many taxa and important changes in faunal compositions. Based on our results, Nerita gastropods do not appear to have suffered from these global environmental changes since their diversity increased rapidly during this period.

Increased availability of suitable habitats and the high spatial and temporal heterogeneity of the environment during Oligocene and early Miocene have likely modified the distribution and connectivity of populations and boosted diversification by increasing the opportunities for allopatric and peripatric speciation. The pattern we found is congruent with that of three other intertidal or shallow marine gastropod genera: Conus, Echinolittorina and Turbo. The observed slowdown of diversification in these genera during late Miocene and Pliocene was interpreted as a consequence of limited speciation opportunities, due to the progressive filling of newly created niches.

Diversification rates in Nerita have likewise decreased slightly since the end of the Oligo-Miocene period. Contrastingly, the N. undata complex (originating from the IAA) presents a higher probability of diversification as shown by the RC test ($\rho = 0.0076$, $\alpha = 0.05$). Originating during mid-Pliocene, this clade has diversified since this period in 11 robust genetic clades: 2 in the SWIO and 9 in the IAA. The detection of different geographically restricted lineages within species with Indo-Pacific distributions, like N. albicillaor N. undata, suggests that dispersal occurs at relatively small geographic scales, despite a high dispersal potential due to long-lived planktotrophic larvae.

The spatial and temporal heterogeneity of the IAA region may enhance species diversification at small geographic scales by constantly modifying connectivity between populations for species with benthic adult and planktonic larval stages which are dependent on ocean currents and available habitats for settlement. Our results, like in other gastropod genera, support the centre-of-origin hypothesis for Nerita, the IAA presenting significantly more diversification events during the Oligocene.

THE PRINCIPLES OF ECOSYSTEM MANAGEMENT

Even if some synergies between the different international arrangements and processes relating to the biodiversity of freshwater resources still deserve more attention and none of the international treaties and institutions already deals fully with freshwater ecosystems, there is an increasing consensus on the highest political levels to manage water, water-related processes, and ecosystems in a sustainable manner. However, at the level of implementation the value of ecosystems is rarely taken fully into account. In order to close this implementation gap the contracting parties of the CBD and other advocates have proposed a so-called "ecosystem-based approach" as an appropriate way of managing natural resources.

The Ecosystem Approach

The ecosystem approach is a strategy for the integrated management of land, water and living resources that promotes conservation and sustainable use in an equitable way. The term ecosystem is used more as a mental construct suggesting complexity and systems interaction rather than a geographic entity. An emphasis on the complexity of system-wide interactions highlights scientific uncertainty and results in the need to deal with this uncertainty explicitly by acting conservatively and managing adaptively: setting a course of action based on a set of hypotheses, monitoring what happens, and re-evaluating the direction on what one learns. At its most fundamental level, an ecosystem approach maintains diversity as a means of building resilience against catastrophic events in biological, economic, organisational, and political systems. At its second meeting in 1995, the Conference of the Parties of the CBD adopted the ecosystem approach as the primary framework for action, and subsequently has referred to the ecosystem approach in the elaboration and implementation of the various thematic and cross-cutting issues work programs under the Convention. The issues concerned include:

- Biological diversity of inland water ecosystems.
- Marine and coastal biological diversity.
- Agricultural biological diversity.
- Forest biological diversity.
- Indicators of biological diversity.
- Incentive measures and environmental impact assessment.

By recommending the ecosystem approach as the guiding concept, in May 2000 the fifth COP took a next step towards its practical verification in the frame of the CBD.

Concerning inland water ecosystems a work programme was adopted by the COP which highlights the importance of designing integrated watershed, catchment and river basin management strategies. Particularly the interrelation between water management and agriculture was stressed, as well as the need to use the ecosystem approach in all sectors which have an impact on inland water biodiversity.

The ecosystem approach was initially developed as a research paradigm and not intended to serve as the basis for resource management. With the political application of the notion "ecosystem-based approach" advocates intended to find an adequate language in response to the shortcomings of classical nature conservation approaches. Although classical nature conservation (e.g. designation of protected areas) can be considered successful in many regions and also is an important module in ecosystem management, its recognition of human needs and other legitimate sectoral interests was rather insufficient. It did not put enough emphasis on the biological diversity outside protected areas. Furthermore, the primary method of protecting freshwater biodiversity has been to designate particular species as threatened or endangered making them subject to national recovery programs or international protection.

Unfortunately, this approach is failing and it demonstrates the need for a more holistic approach in order to overcome the shortcomings of traditional efforts in nature protection policy. In the United States, for example, no aquatic species has ever graduated from the government's endangered species list, but 10 species of fish have been removed due to extinction. Equally, ecosystem management responds to the obvious shortcomings of classical single-resource approaches focused on exploitation.

The 12 Malawi Principels

- Management objectives are a matter of societal choice.
- Management should be decentralised to the lowest appropriate level.
- Ecosystem managers should consider the effects (actual or potential) of their activities on adjacent and other ecosystems.
- Recognising potential gains from management there is a need to understand the ecosystem in an economic context. Any ecosystem management programme should:
  - Reduce those market distortions that adversely affect biological diversity.
  - Align incentives to promote sustainable use.
  - Internalise costs and benefits in the given ecosystem to the extent feasible.
- A key feature of the ecosystem approach includes conservation of ecosystem structure and functioning.
- Ecosystems must be managed within the limits to their functioning.
- The ecosystem approach should be undertaken at the appropriate scale.

- Recognising the varying temporal scales and lag effects which characterise ecosystem processes, objectives for ecosystem management should be set for the long term.
- Management must recognise that change is inevitable.
- The ecosystem approach should seek the appropriate balance between conservation and use of biological diversity.
- The ecosystem approach should consider all forms of relevant information, including scientific and indigenous and local knowledge, innovations and practices.
- The ecosystem approach should involve all relevant sectors of society and scientific disciplines. These principles were agreed at CBD's COP-5 under decision V/6 with minor modifications.

Therefore, ecosystem management is proposed as a modern way of managing natural systems and generally intends to use natural resources within their natural limits. However, although there are still some controversies on how to translate ecosystem management into concrete management terms, parties of the CBD have already agreed on basic principles to guide the *modus operandi*.

A key point is that the ecosystem approach is a holistic process for integrating and delivering in a balanced way the three objectives of the CBD: conservation and sustainable use of biological diversity combined with equitable sharing of biodiversity's benefits. In particular, the agreed principles of the ecosystem approach stress the importance of:

- Societal choice concerning management objectives.
- The "economics" of biodiversity (internalisation of costs and benefits, balance between conservation and use).
- Adequate institutional structures (decentralisation, management at the appropriate scale).
- Adaptive management which enables policy makers to anticipate and cater for the natural dynamics of ecosystems.
- Consideration not only of scientific knowledge but also of indigenous and local knowledge.

In order to translate this rather abstract principles into guidelines for water management it is first necessary to define the appropriate management level to meet the general objectives. While the cited principles stress the importance of decentralised approaches in general terms, the idea of river basin management is considered an adequate starting point for the implementation of the ecosystem approach in water management. The reasoning is that river basins offer many advantages for strategic planning and that they represent the most appropriate geomorphologic units on which to base ecosystem management practices. However, concerning ecosystem protection there are some limits: groundwater catchments only rarely match with surface water catchments; water quality and quantity inside the river basin interacts with other influences outside the catchments boundaries; estuarine ecosystems, for example, can only be sufficiently protected by taking into account the interaction of freshwater and tidal currents.

Therefore, river basin management should not be perceived as a panacea but it does provide a sound geographical basis for implementation. Briefly, ecosystem management should be an integral component of comprehensive river basin management strategies. Hence, the explicit switch to ecosystem management leads to an accentuation of the main focus in water management because ecosystem management demands:

- A broader perspective: whole ecosystems and the full range of services, no sectoral view.
- Efforts to balance natural ecosystem robustness and human-induced alterations.
- Collation of information on the status of, and threats to, the biodiversity of the freshwater ecosystem concerned (including impacts of the introduction of alien species).
- Explicit consideration of land-use, in-stream flow and water temperature for maintaining ecological functions.
- A long-term perspective.
- Consideration of the full economic value of freshwater ecosystems (include its biodiversity).
- Recognition of ecosystem's change as a matter of fact (i.e. restoration of damaged or destroyed ecosystems will never lead to the "pristine" state; adaptive management to deal with uncertainties).
- Integration of scientific knowledge and modern management tools (Decision Support Systems, Geographical Information Systems etc.) into decision-making.
- Identification of win-win options between protection and use of freshwater ecosystems: "use them, but don't lose them".
- Consideration of the full range of policy instruments applicable in water management (economic and legal instrument, co-operation etc.)

As most of these elements are widely agreed upon the rhetoric of international water policy, the next steps must include attempts to tackle the implications of the ecosystem approach in practice. For example, only little is known about the value of ecosystem management in political terms for at least two reasons:

- The concept appears highly demanding because much information – sometimes on "invisible" relationships between human-induced pressures and natural reactions – is needed, and effective vertical and horizontal co-ordination of sector policies and planning instruments has to take place. Today, even in wealthy nations a lack of integration across sectors in managing water resources can be identified.
- It is necessary to decide which services of water receive priority, and how the costs and benefits are to be distributed. The developed world, for example, can "afford" to stop or to remove development projects in favour of ecosystem integrity (e.g. Great Whale Hydropower Project in Canada, Edwards Dam in the US) and to use sometimes generous subsidies to compensate land owners for restricting their use of the lands. The restoration of river ecosystems in the developed world also is an example for costs emerging in the short terms. Although long-term economic benefits are regularly provided, political decision are often dominated by short-term considerations.

ISOLATION BY DISTANCE AS A DRIVING PROCESS OF DIVERSIFICATION IN THE INDIAN OCEAN

The genus Nerita being almost restricted to tropical rocky shores, its distribution is partly correlated to the existence of these particular habitats. Although the cryptic diversity of IAANerita can be explained by global climatic variations and environmental modifications over geological time, these factors do not explain the high proportion of endemic and cryptic lineages found within the SWIO. In this region, the volcanic activity (geological hotspot) started more than 65 Ma ago and created a North-South oriented chain of islands across the Indian Ocean: Laccadive islands (emergence: 65-60 Ma), Maldivian islands (60-50 Ma), Chagos archipelago (50-49 Ma), Mascarene plateau (48-31 Ma), Mauritius island (8 Ma) and finally Reunion Island (2 Ma).

During sea level low stands, these islands represented large landmasses, particularly during the Miocene. Various terrestrial clades used those multiple islands as stepping-stones to colonize the SWIO, while subsequent sea level rises facilitated secondary isolation and speciation. Thus, colonization events of SWIO by Asian and IAA species favoured terrestrial speciation due to the action of geologic/climatic events throughout this period.

This model seems to apply to Indian Ocean Nerita species as well. Our hypothesis is supported by the old asynchronous divergence, ranging from 21 to 5 Ma (the Mascarene plateau emerged during this period) of three endemic species from their closest parent species living in the IAA region: N. aterrima, N. magdalenae, N. umlaasiana. Nowadays, there is little connectivity between the western and eastern Indian Ocean populations for a wide range of marine taxa. However, throughout the Miocene and assuming no major changes in ocean currents compared to nowadays, the intermittent emergence of volcanic landmasses and new rocky shores in SWIO may have permitted larval colonization from IAA Nerita populations.

Without constant larval input due to sea level variations changing distances between populations of IAA and SWIO, newly settled populations diverged and formed new species by allopatry or peripatry. Ecological transition has not played a role in the formation of SWIONerita species as all sister species-pairs occupy the same ecological niches: lower littoral for N. albicilla-N. sanguinolenta; mid-littoral for N. textilis-N. exuvia; upper littoral for the pairs N. magdalenae-N. costata and N. quadricolor-N. spengleriana; supra-littoral for N. insculpta-N. umlaasiana. Ecological conservatism have been identified in other intertidal gastropods, e.g., sister species of Echinolittorina remain allopatric for millions of years without changing their habitat preferences. Ecological conservatism during diversification has also been documented in other taxa, such as coral reef fishes. For a temperate terrestrial gastropod (Arion subfuscus), the habitat fidelity over time (as evidenced by the persistence of allopatry) has even contributed to the increase of lineage accumulation during the past glacial maximum.

Therefore SWIO Nerita endemics followed a "terrestrial" diversification pattern in the region and formed due to the synergy of several abiotic factors: the presence of an active geological hotspot and sea level variations, favouring colonization of Nerita populations from the IAA and subsequent genetic isolation. Changes in ocean circulation may have also played a role, but modelling currents at small geographical scale through geological time seems presently not feasible.

GEOGRAPHICAL DISTRIBUTION OF ECOSYSTEMS

The global distribution of plant communities generally tracks global latitudinal climate zones and the elevation-dependent patterns of temperature and precipitation in alpine topography. Five broadly defined vegetation zones — semi-arid grasslands, temperate forests, tropical forests, arid deserts, and polar regions — characterize the global distribution of plant communities. Each zone has distinctive ground surface coverage, reinforcement, soil types, and weathering properties that result from its plant communities. Plants in grassland and forest zones not only generate different types of soils, but their root systems reinforce soils differently so the landscapes differ in their resilience to environmental disturbance as well. Grasslands generally have more biomass below ground than above ground, most of it in roots.

Grasslands are particularly vulnerable to erosion following overgrazing and plowing. Temperate forests can be deciduous or coniferous. In temperate forests, extensive root networks form interlocking webs that mirror the extent of the forest canopy and significantly reinforce hillside soils. Geomorphically important effects of forest type include the depth and strength of root penetration, and the shape of fallen trees that enter rivers where the difference between a long pole shape typical of conifers and the branching structure typical of deciduous trees influences the stability and transportability of wood. Tropical forests typically have little below ground organic matter and extensively weathered soils. As plant nutrients are held in the plants themselves, it can be hard to re-establish native forests after forest clearing. In contrast to temperate and tropical regions where vegetation can have a major impact on the type, frequency, and intensity of geomorphological processes, vegetation plays a relatively minor geomorphological role in arid and polar landscapes. However, even in desert landscapes where plant communities have little biomass or plant cover, the presence or absence of even a little ground cover or a thin web of roots can greatly affect soil development. And frozen soils that support small trees and tundra vegetation can accumulate and store lots of soil carbon due to slow breakdown. The potential for warming tundra soils to release carbon into the atmosphere is one of the great concerns surrounding potential feedbacks between climate change and ecosystem response.

The distribution of animal communities also generally tracks climate zones. Penguins don't live in deserts, and camels don't live on glaciers. Although most animals only marginally influence the landscapes they inhabit, a few types of animals have a substantial geomorphological influence. For example, mass spawning salmon move up to half the gravel transported by some rivers in the Pacific Northwest. Overgrazing by domestic animals accelerates soil erosion and can trigger gully development. Burrowing animals and mound building ants and termites can displace and mix tremendous amounts of soil and weathered rock. Charles Darwin calculated that over the course of centuries worms steadily plowed the hillside soils of England. Plants and animals influence geomorphological processes directly, as when burrowing activity and roots mechanically pry rocks, and indirectly, as when plants protect soils from erosion during precipitation events, and roots mechanically reinforce slope-forming materials. Plants also are central to the chemical transformations that accompany the breakdown of rock-forming minerals into clay minerals that hold nutrients essential to soil fertility. Plant communities thus shape how a landscape weathers, erodes, and supports life. The size of an organism is not necessarily related to its importance or impact. Soil bacteria, for example, are essential chemical weathering agents in many environments.

Humans

Humans are today the most widely distributed species on the planet, and we move enough rock and soil to count among the primary geomorphic forces shaping Earth's modern surface. Coal and mineral mining operations move whole mountains and excavate great pits, farmers' plows push soil gradually but persistently downhill, and construction crews cut or fill the land to facilitate building or to suit our aesthetic whims. Human activities further influence geomorphological processes through the indirect effects of our resource management and land use practices. In manipulating our world, we alter hydrological processes by changing surface run-off, stream flow, and flood flows. Clearing stabilizing vegetation and changing water fluxes in the landscape affects slope stability, and increases erosion rates, and construction of dams and coastal jetties interupts the transport and storage of sediment. Such changes often have unintended consequences far downstream, as when upriver dam construction starves beaches and deltas of sand and mud by trapping the sediment that formerly nourished coastal environments. Learning to recognize and understand such connections is central to applied geomorphology, whether to aid in the design resilient communities, develop more sustainable land use practices, or construct measures to protect critical infrastructure. Through history human activities have resulted in changes to a wide range of local and landscape-scale geomorphological processes that we are still learning to recognize and fully appreciate. For example, we are just coming to understand the profound impact of Bronze Age forest clearing on European rivers. Further dramatic changes in regional and global land cover are projected in the coming century. The influence of human activity is already great enough on a global scale that geologists have proposed we are entering a new era of geologic time that they call the Anthropocene (the human era). Over the next century, changes in the global climate are predicted to cause increasingly variable weather, more frequent hurricanes, rising sea levels, and a host of related regional impacts, like the loss of winter snow pack in the Pacific Northwest. Predicting the ways that landscapes respond to such changes will be central to planning societal adaptation or mitigation efforts.

Landscapes

Landscapes are suites of contiguous landforms that share a common genesis, location, and history. The study of landscapes involves investigations over a tremendous range of spatial and temporal scales, from the mobilisation of a single grain of sand from a river bed over a few seconds, to the rise of the Himalaya and growth of the Tibetan Plateau over millions of years. Consequently, the approach, methods, and scale of analysis involved in any particular study need to be tailored to the questions the geomorphologist seeks to address. Landscapes can be divided into distinct units that can be studied over discrete periods of time, but it is the relationships between processes and landforms that are truly fundamental in our modern approach to geomorphology. In seeking to understand landscape evolution it is important to appropriately match measurement and understanding of geomorphological processes to the spatial and temporal scales over which relevant processes act.

Process and Form

Landscape forms and surface processes are inseparably linked. Stream flow, slope failure, flowing glaciers, and blowing wind act to shape landscapes. At the same time, topography itself determines the style and rate of geomorphological processes. The shapes and orientations of larger-scale landforms generally control the rates and distributions of the small-scale erosional and depositional processes that determine how landforms evolve over time. The fundamental interdependence of process and response means that a geomorphologist can often read form to infer process, as in determining dominant wind directions from the shape and orientation of sand dunes. But in order to understand landforms and predict landscape response, we must understand the processes that form them. Consequently, the relationship between process and form lies at the heart of geomorphology.

The geological and environmental history of a region can leave a lasting signature on landforms and on the processes operating upon them.

The physiographic signature of long-vanished glaciers still dominates the topography of alpine regions around the world. It should not be surprising then that process geomorphology and historical geomorphology are complementary disciplines. Although many geomorphologists specialize in one or the other approach, no geomorphologist can afford to ignore either. Proposing testable hypotheses and interpreting field, laboratory, and simulation data in the evaluation of those hypotheses requires consideration of processes over time because of the fundamental linkages between landscape history, process, and form.

Landscape dynamics and evolution result from the interaction of multiple systems that can be studied and understood in their own right. The topography of a mountain belt reflects the interaction of tectonic processes that raise rocks above sea level with the hydrologic and run-off processes that govern how erosion acts to shape slopes and incise valleys. Considered broadly, the interaction of coupled tectonic, climatic, and erosional systems control patterns of uplift, erosion, and sedimentation as they change over geologic time. At finer scales, many geomorphological processes reflect the interaction of ecological and hydrological systems, as when the binding effect of tree roots helps to stabilize soils on landslide-prone slopes, or when log jams create dams that divert stream channels to new courses across their floodplains. Understanding landscapes often requires an a ppreciation of how systems interact in different regional contexts and environmental settings.

Spatial Scales

At global and continental scales, geomorphologists study major physiographic features like mountain belts, depositional basins, and great river systems like the Mississippi, Amazon, or Nile. At these scales, broad patterns in global climate and plate tectonics influence patterns of erosion and deposition that, in turn, influence the size and extent of mountains, plateaus, lowlands, coastal plains, and river basins. Such patterns also affect the distribution of soil types and the relative importance of glacial, fluvial, eolian (wind-driven), and coastal processes in shaping topography.

At a regional scale, distinct physiographic provinces reflect areas with broadly similar history, landforms, and landscape dynamics. Examples of such provinces include mountain belts like the Sierra Nevada or Appalachians as well as features like the Great Plains, California's Central Valley, and the Colorado Plateau in the American Southwest. Physiographic provinces are areas in which similar suites of geomorphological processes govern landscape formation and dynamics, and thus where one finds similar suites of landforms. A drainage basin is the land surface area drained by a given stream. Small streams come together to form larger rivers, so landscapes are naturally organized into smaller drainage basins nested within larger drainage basins.

Drainage basins range in size from a headwater catchment that collects water from a single mountain hillside to the Amazon River basin that drains more than half of South America. Drainage divides are topographic ridges that separate drainage basins.

Because streamflow is the predominant agent of erosion and sediment transport on land, drainage basins are the logical unit for analysis of many geomorphological processes. At the scale of individual valley segments, the tectonic or climatic setting of a region often defines areas where different types of processes and/or histories have led to development of distinct landforms and dynamics. The difference between U-shaped valleys carved by glaciers valleys and V-shaped valleys cut by streams is a classic example. Systematic downstream changes in stream valley morphology, from erosional headwater streams with narrow valleys that are confined between bedrock walls, to broad unconfined valleys of depositional lowlands. Distinct suites of valley segment types are diagnostic of specific physiographic provinces, and their character and distribution both vary regionally and generally reflect the history and processes of landscape evolution and shape ecosystem dynamics.

At finer spatial scales, landscapes can be divided into distinct hillslopes, hollows, channels, floodplains, and estuaries. Hillslopes (including hilltops) are the undissected uplands between valleys. Hollows are unchanneled valleys that typically occur at the head of channels in soil-mantled terrain. Channels are zones of concentrated flow and sediment transport within well-defined banks, and floodplains are the flat valley bottoms along river valleys that are inundated during times of high discharge under the present climate. Estuaries are locations where streams enter coastal waters to arrive at their ultimate destination — sea level.

Temporal Scales

The scale of observations in both time and space strongly influence geomorphic interpretations. Matching the scale of observation to the scale of the question you seek to answer is critical for gathering meaningful data. It's easy to understand the importance of spatial scale. Measuring erosion on a single hillside in Kenya won't tell you much about the erosion rate of the African continent. It is more difficult to understand the influence of time on geomorphic data, observations, and interpretations.

Consider how the first estimates of continental erosion rates, which were made by measuring the concentration of suspended sediment exported by rivers over a few years, turned out to be wrong. While this is a reasonable approach if the measurements capture the variability of the system over time, in this case they did not. With only a few years of data, sediment yield results are likely to be biased by the short period of observation because most streams systems experience rare but massive floods that periodically transport immense amounts of sediment. How would one know that the few years in which the sampling occurred are representative of a meaningful, long-term average? The issue of how to integrate the influence of large and small events in shaping larger landforms remains central to modern process geomorphology.

Topography evolves over periods of time that range from the millions of years that are required to erode away mountain belts, to the few seconds, minutes, or days it takes for a landslide, flood, or earthquake-driven fault displacement to disrupt the land surface. Climate cycles influence topography over millennia as glaciers advance, retreat, and scour out alpine valleys. Likewise, river profiles adjust to the sea level changes that accompany glaciations. Landscape responses to large-scale disturbances like hurricanes and volcanic eruptions are often evident for centuries, and it can take decades for landslide scars to revegetate and river channels to process the sediment shed from slopes during large storms. River flow exhibits annual and seasonal variability that controls the timing of sediment movement and structure of stream ecosystems. Because of the disproportionate influence of infrequent extreme events like storms, landslides, and floods, rates of processes measured over short time spans may not adequately characterize average rates over longer time scales. Not surprisingly, geomorphologists deal with a wide variety of measurements over different time scales, from rates directly measured in the field, to indirect measurements of long term erosion rates inferred from isotopic analyses, and erosion rates constrained by sedimentary volumes preserved in depositional basins. The key, of course, is to employ analyses relevant to the time scale of interest.

3

Biogeography and Genesis of Soils

UNDERSTANDING THE BIOGEOGRAPHY

Biogeography is a branch of geography that studies the past and present distribution of the world's many species. It is usually considered to be a part of physical geography as it often relates to the examination of the physical environment and how it affects species and shaped their distribution across space. As such it studies the world's biomes and taxonomy—the naming of species. In addition, biogeography has strong ties to biology, ecology, evolution studies, climatology, and soil science.

History of Biogeography

The study of biogeography gained popularity with the work of Alfred Russel Wallace in the mid-to-late 19th Century. Wallace, originally from England, was a naturalist, explorer, geographer, anthropologist, and biologist. He first extensively studied the Amazon River and then the Malay Archipelago. During his time there, he examined the flora and fauna and came up with the Wallace Line—a line that divides Indonesia apart and the distribution the animals found there. Those closer to Asia were said to be more related to Asian animals while those close to Australia were more related to the Australian animals. Because of his extensive early research, Wallace is often called the "Father of Biogeography."

Following Wallace were a number of other biogeographers who also studied the distribution of species. Most of those researchers looked at history for explanations, thus making it a descriptive field. In 1967 though, Robert MacArthur and E.O. Wilson published The Theory of Island Biogeography. Their book changed the way biogeographers looked at species and made the study of the environmental features of that time important to understanding their spatial patterns. As a result, island biogeography and the fragmentation of habitats caused by islands became popular as it was easy to explain plant and animal patterns on islands. The study of habitat fragmentation in biogeography then led to the development of conservation biology and landscape ecology.

TYPES OF BIOGEOGRAPHY

Today, biogeography is broken into three main fields of study. The three fields are historical biogeography, ecological biogeography, and conservation biogeography. Each field, however, looks at phytogeography and zoogeography.

Historical biogeography is called paleobiogeography and studies the past distributions of species. It looks at their evolutionary history and things like past climate change to determine why a certain species may have developed in a particular area. For example, the historical approach would say there are more species in the tropics than at high latitudes because the tropics experienced less severe climate change during glacial periods. This led to fewer extinctions and more stable populations over time.

The branch of historical biogeography is called paleobiogeography because it often includes paleogeographic ideas-most notably plate tectonics. This type of research uses fossils to show the movement of species across space via moving continental plates. Paleobiogeography also takes varying climate as a result of the physical land being in different places into account for the presence of different plants and animals.

Ecological biogeography looks at the current factors responsible for the distribution of plants and animals. The most common fields of research within ecological biogeography are climatic equability, primary productivity, and habitat heterogeneity.

Climatic equability looks at the variation between daily and annual temperatures. It is harder to survive in areas with high variation between day and night and seasonal temperatures. Because of this, there are fewer species at high latitudes because more adaptations are needed to be able to survive there. In contrast, the tropics have a steadier climate with fewer variations in temperature. This means plants do not need to spend their energy on being dormant and then regenerating their leaves and/or flowers, they don't need a flowering season, and they do not need to adapt to extreme hot or cold conditions. Primary productivity looks at the evapotranspiration rates of plants. Where evapotranspiration is high, so is plant growth. Therefore, areas like the tropics that are warm and moist foster plant transpiration allowing more plants to grow there. In high latitudes, it is simply too cold for the atmosphere to hold enough water vapour to produce high rates of evapotranspiration and there are fewer plants present.

Finally, habitat heterogeneity leads to the presence of more biodiversity. After looking at the various fields in historic and ecological biogeography, conservation biogeography developed. This is the protection and/or restoration of nature and its flora and fauna.

Biogeography is important as a branch of geography that sheds light on the natural habitats around the world. It is also essential in understanding why species are in their present locations and in developing protecting the world's natural habitats.

GENESIS OF SOIL

Soils of the coastal areas present generally little evolution, since they are affected by erosional-depositional events, oscillating water-table, spatial variability of texture, carbonate and organic matter content. Leaching, decarbonation, brunification, gleyzation have been recognised as the most active soil forming processes in these areas in temperate regions. Also anthropic intervention contributes in modifying soil development: sand and water extraction, terrain levelling, tourism enhancing, land use changing, all of these contribute to new environmental conditions that may affect pedogenesis. Correspondingly, the natural vegetation of these areas may be subjected to change with changing of environmental conditions.

Similar modifications have been recorded recently in coastal and wetland areas of North-East Italy, where land reclamation and changes in management in the last 100 years determined new conditions for the soil genesis and the development of the vegetation cover.

The objectives of this work were:

- To examine the soil distribution in these sensitive areas, which constitute examples of pedosites subject to disappear with changing of environmental conditions;
- To relate soils and phytocoenoses with peculiar ecological characteristics; and
- To indicate a trend of pedogenesis which might be applied to areas subjected to watertable or sea level variations, in consequence of hypothetical climatic changes and in relation to their future management.

Materials and Methods

Site Location

The investigated area is part of the perilagoonal belt located in the northeastern part of Italy, between the river Isonzo and the Venice lagoon. Geologically, it is composed of Holocene mainly sandy alluvial materials. All over the Holocene, the coastline was close to the present. However, subsidence, eustatism and variations in river-transported solid materials determined a peculiar topography, with depressed wetlands and hydromorphic areas alternating with drier ones, located in more elevated places.

Data elaboration on the basis of the bioclimatic indexes proposed by Rivas-Martinez et al. shows that the present climate in the study area is temperate submediterranean subhumid. The mean annual temperature is 14°C. Mean annual precipitation ranges from about 950mm in the southern part and 1,035mm in the northern part. The Soil Taxonomy, the calculated Soil Temperature Regime is mesic; and Soil Moisture Regime is variable from aquic to udic-xeric depending on local topographical conditions.

The vegetation of the area is characterised by the mutual influence of different chorologic elements: C-European, Mediterranean, Orophilous and Illyric. Such a compenetration, together with the peculiar environmental conditions, is responsible for the coexistence of plant communities with both a C-European and Mediterranean trend. At the same time, the local topography determines the presence of the edaphoxerophilous series at some sites, and the edaphohygrophilous one at other ones.

Among the different phytocoenoses in the investigated area, the most interesting are the hygrophilous natural and semi-natural fens and meadows, like the endemics Erucastro-Schoenetum nigricantis and Plantagini altissimae-Molinietum caeruleae, growing on neutral to subalkaline soils enriched in organic matter. The latter association is common in areas where water-table is close to the surface, and the continuous agricultural management enhances preservation of these prairies. Progressive abandonment determines littering, auto-manuring and development of common shrubs, and successively wood vegetation.

CHEMICAL WEATHERING AND SOIL

Chemical weathering occurs because the minerals in rocks form at deep Earth pressure and temperature conditions that are not in equilibrium with conditions at Earth's surface, and are thus vulnerable to chemical decomposition and transformation. A primary weathering agent is rainwater that percolates into the ground and promotes chemical weathering because it contains dissolved ions gained from the atmosphere and from the soils through which it moves. The bipolar water molecule is a potent solvent — given time. The metabolism of soil microorganisms and decay of organic matter enhance weathering as they add organic acids to water moving through soils. Root respiration and microbial oxidation make a soil atmosphere rich in CO_2. The addition of water makes carbonic acid. Sulfuric and nitric-acid weathering are important in some areas and always present at some level. In addition, biochemical activity tends to increase rates of chemical reactions between soil fluids and minerals by lowering pH and increasing temperature.

Processes of chemical weathering include solution, oxidation and reduction, hydrolysis, ion exchange, and the formation of new, more stable secondary minerals, like clays and hydrous oxides. The end results of chemical weathering depend on a variety of interacting factors including the composition and texture of the parent material and the chemical, physical, and biochemical processes acting in a particular environment. The mobility and stability of the secondary minerals and solutions produced depend on environmental conditions like pH, Eh, and temperature. Chemical weathering is the breaking of chemical bonds — metallic, ionic, and covalent. The corresponding principle weathering processes are electron exchange (oxidation/reduction), ionisation (solution), and ion exchange (as in acid attack). As many rocks are dominated by a mix of ionic and covalent bonds, solution and acid attack are major weathering processes. Hydration and dehydration are also important weathering mechanisms for certain rock types and in certain environments. Chemical weathering is essential for the biosphere in the critical zone, where vegetation demand for Ca, Mg, K, NO_3, and P is extraordinary. Nutrients derived from minerals are cycled through ecological systems because of the slow pace of weathering or the depleted nature of surficial materials (especially in key limiting elements such as phosphorus).

Oxidation / Reduction

Oxidation is a process during which an element loses an electron to a receptor, often an oxygen ion, like when iron rusts. Conversely, reduction is defined as the gain of an electron. Free oxygen is rare at crustal depths where rocks form, but abundant at Earth's surface. Rocks and rock-forming minerals typically oxidize when they are exposed to well-oxygenated soil water, directly to the atmosphere, or to gases in soil pores. Reducing conditions in oxygen-poor waters with lots of organic matter, like swamps and peat bogs with high seasonal water tables, generally prevent oxidation, retard organic decay, and slow down weathering. When soils alternate between saturated and unsaturated conditions, a speckled colour pattern known as mottling develops — with gray colours due to the lack of oxidized iron as well as reddish colours due to oxidation.

Oxidizing potential is expressed in terms of redox potential (Eh), the availability of free oxygen, which is greatly influenced by the amount of dissolved organic matter in pore fluids. Soils typically have Eh high enough to oxidize most common elements, but iron, manganese and sulfur are especially prone to rapid oxidation and typically occur as red, black, and yellow coatings in soils. Redox potential exerts a substantial influence on ion mobility, and oxidation is often the first form of weathering to alter freshly exposed rock surfaces. Oxidized versions of elements are relatively immobile, whereas reduced versions are far more mobile. Over time, rinds of oxidized material form on the surfaces of outcrops or boulders as other material is leached away. When rocks containing common iron bearing carbonates, sulfides, and silicates (such as olivine and biotite) oxidize, they become susceptible to additional physical weathering. Oxidation produces relatively insoluble ferric oxides, like hematite (Fe_2O_3), or rust, and oxyhdroxides like goethite (FeO(OH)) that colour soils and weathered rock various shades of reddish or yellowish brown. By the same token, oxygen starved conditions (such as those under stagnant water rich in decomposing organic matter) turns iron and manganese back into reduced forms, which allows them to be dissolved and leached if the water drains or is flushed from the soil.

Solution

The flow rate and acidity of pore water are two of the most important factors influencing the amount of disolution from soil, sediment, or rock in a given weathering environment. In addition, pH strongly affects the solubility of most elements. Rainwater is slightly acidic from dissolved atmospheric CO_2, and chemical and biologic weathering processes often act to lower the pH of water moving through a weathering zone. In weathering zones with active groundwater circulation, fresh water comes in contact with parent material, and weathering continues as leaching removes dissolved material. Slowly circulating pore waters retard dissolution as the number of dissolved ions in solution approaches an equilibrium concentration.

Natural groundwater tends to be slightly acidic due to dissolution of carbon dioxide in water to produce carbonic acid ($H_2CO_3^-$). Carbonic acid is not a strong acid but it is extremely abundant because it forms wherever water encounters CO_2 through the carbonation reaction:

$$H_2O + CO_2 \longleftrightarrow 2H_2CO_3^-$$

Decay of organic matter together with respiration of soil invertebrates, bacteria, and root systems can elevate CO_2 concentrations in soil pores so that they are 10 to 100 times greater than atmospheric concentrations. This makes carbonation a particularly important factor in heavily vegetated areas. Cold temperatures also favour formation of carbonic acid in soil water, because the solubility of CO_2 is inversely proportional to temperature, as is true of most gases.

In solution, carbonic acid readily disassociates into hydrogen (H^+) and bicarbonate ions (HCO_3^-):

$$2H_2CO_3^- \longleftrightarrow H^+ + HCO_3^-$$

Consequently, bicarbonate is the most common cation in natural groundwater. Atoms exposed on the mineral surfaces of rock and soil particles are electrically charged ions that react with dissociated hydrogen (H^+) and hydroxide (OH^-) ions in water. This interaction breaks bonds, effectively disassociating individual mineral molecules and causing exchanges that release cations from the mineral surface into solution. Mineral structures become unstable and vulnerable to further weathering when they lose cations, so a little weathering promotes more weathering. Congruent dissolution occurs when all the constituents of an individual molecule are separated and remain in solution. During incongruent dissolution, some of the released ions recombine to create new compounds and secondary minerals. Dissolved material may remain in solution and move along with flowing water and re-precipitate elsewhere, or it may emerge into streams and rivers, and eventually reach the ocean. The origin of the salinity of the oceans lies in the long-term delivery of dissolved material in stream water. Most common elements are soluble to some degree in both rainwater and soil water. Consequently, water circulation promotes solution by introducing fresh water that removes dissolved ions from mineral surfaces.

The dissolution of calcite ($CaCO_3$, calcium carbonate) is a particularly important chemical weathering reaction. This occurs in the presence of carbon dioxide (CO_2) and introduces bicarbonate ions (HCO_3^-) into solution. The resulting reaction is expressed as

$$CaCO_3 + H_2O + CO_2 \longleftrightarrow Ca^{++} + 2HCO_3^-$$

The carbonate dissolution reaction is reversible. An increase in CO_2 concentration within soil gasses, a decrease in pH, or dilution will drive the reaction to the right (as written above); the rate of carbonate dissolution will increase and the bicarbonate concentration in groundwater will go up. This effect helps percolating water erode fractures and form extensive cave systems typical of regions underlain by carbonate rocks (limestone or dolomite). Conversely, decreased CO_2 concentration, increased pH, or evaporation will drive the reaction to the left and favour precipitation of calcium carbonate ($CaCO_3$). It is this reaction that deposits stalagmites and stalactites in caves, as well as calcite in desert soils.

When carbonic acid dissociates to form an "acid" of protons, the resulting weathering of aluminosilicate minerals consumes CO_2 and thus helps to cool global climate through the general reaction

$$\text{aluminosilicate} + H_2O + CO_2 \rightarrow \text{clay mineral} + \text{cations} + OH^- + HCO_3^- + H_4SiO_4$$

Earth's long-term climate is thus mediated by organic matter burial and carbonate formation (both of which sequester carbon in the geologic record) and silicate weathering, which consumes CO_2 (producing bicarbonate). Over the long run, glaciations and the anthropogenic contribution to atmospheric CO_2 are short-term perturbations of this geologic control on global climate through carbonate burial and the influence of weathering on the concentration of CO_2 in atmosphere.

Calcite and salts are readily dissolved in water, so carbonate rocks and evaporites are particularly susceptible to dissolution, especially in regions with abundant precipitation.

In contrast, quartz and most other rock-forming silicate minerals are not very soluble at typical Earth surface conditions, leading to slow rates of dissolution in most environments.

Solubility varies greatly between minerals, but even the least soluble minerals will dissolve over time if exposed to solutions refreshed by water circulation.

Iron and aluminum oxides, however, are virtually insoluble under oxygenated soil water conditions, so these compounds are typically left behind as more soluble, mobile material is depleted. Consequently, the abundance of iron and aluminum oxides increases as rocks and sediment are exposed to weathering. The bright red soils of the southeastern United States are an example of how these oxides can accumulate through time. Dissolution can play an important role in increasing pore space and thus increasing the percolation of water, soil acids, oxygen, and bacteria into the regolith.

Mineral Stability

The general susceptibility of rock minerals to weathering is the inverse of the sequence in which they form deep within the earth. Rocks that formed at the greatest temperatures and pressures are furthest from equilibrium at surface conditions and are therefore most susceptible to weathering when exposed to the elements.

Among the common silicate minerals that account for the majority of rock forming minerals, olivine and pyroxene are most susceptible to weathering, followed in order of decreasing susceptibility to breakdown by amphibole, biotite, feldspar, muscovite and quartz.

This progression, known as Goldich's Weathering Series, is the opposite of the order of crystallisation as magma cools, familiar to geologists as Bowen's Reaction Series. Under similar environmental conditions, rocks composed of more mafic iron- and magnesium-rich minerals (olivine, pyroxene, amphibole, and biotite) will weather faster than those composed of more felsic minerals (feldspar, muscovite, and quartz).

However, it is not always as simple as this. Due to the importance of both covalent and ionic bonds, silicates with complicated mineral structures break down more readily than do those with simpler structures like quartz (SiO_2) or zircon ($ZrSiO_4$), a very stable mineral even though it has a very high melting temperature. So both complexity and formation conditions are central to mineral stability.

The mobility of cations in rock-forming minerals varies greatly, and influences the relative ease and order in which weathering strips cations from rocks and secondary minerals, with the sequence from most to least mobile proceeding as Ca^{2+}, Na^+, $Mg^{2+} > K^+ > Fe^{2+} > Si^{3+} > Fe^{3+} > Al^{3+}$. The most mobile cations (Ca^{2+}, Na^+, Mg^{2+}) are readily stripped from mineral surfaces, tend to remain in solution, and are the first to be lost from rocks as they weather. The least mobile cations (Si^{3+}, Fe^{3+}, and Al^{3+}) are relatively insoluble and become concentrated in residual soils over time as weathering strips away more mobile elements.

Chemical Reaction of Hydrolysis

Hydrolysis is a chemical reaction in which water molecules (H_2O) are split into protons (H^+) and hydroxide anions (OH^-) that react with primary rock-forming minerals to form new compounds (secondary minerals). Hydrolysis is an important chemcial weathering process that acts to break rocks apart and transform silicate minerals into weathering products.

In hydrolysis reactions, mineral cations are released into solution and replaced by hydrogen (H^+), producing a new mineral. This process results in the transformation of aluminosilicate minerals, like feldspars and micas, into various clay minerals. *The conversion of potassium feldspar (orthoclase) into illite or kaolinite clay are examples of hydrolysis:*

Orthoclase (feldspar) illite (clay)

$$3KAlSi_3O_8 + 2H^+ + 12H_2O \rightarrow 2K^+ + KAl_3Si_3O_{10}(OH)_2 + 6H_4SiO_4$$

Orthoclase (feldspar) kaolinite (clay)

$$2KAlSi_3O_8 + 2H^+ + 9H_2O \rightarrow 2K^+ + H_4Al_2Si_2O_9 + 4H_4SiO_4$$

Hydrolysis is not reversible. Once secondary minerals are formed, further weathering can strip additional cations and can convert secondary aluminosilicates like illite into other, more cation-depleted clays. Upon more intensive weathering, each step in the weathering of clay minerals strips additional cations from the mineral structures. Eventually, intensive weathering can leave kaolinite, which consists of just hydrogen, aluminum, silica, and oxygen, and has no additional cations left to exchange. Progressive alteration of silicate minerals due to weathering reduces the complexity of mineral structures.

Process of Hydration

Hydration describes the process in which silicate minerals combine with water or hydroxide ions (OH^-) to form hydrated compounds. Hydration is another way that primary minerals are converted to secondary minerals. Common forms of hydration reactions include the conversion of anhydrite ($CaSO_4$) to gypsum ($CaSO_4 \cdot 2H_2O$), and the formation of relatively insoluble iron and aluminum hydrous oxides, like limonite ($FeO(OH) \cdot nH_2O$) and gibbsite ($AlOH_3$) in regions of intense tropical weathering like the Amazon basin.

Clay Formation

Clay minerals are both a product and a player in processes of hydrolysis and hydration. Unlike most primary minerals, with the exception of quartz, secondary minerals like clays and hydrous oxides are chemically stable under earth surface conditions. They become a major constituent in soils because their relative stability and immobility leaves them as common in situ byproducts of weathering.

Most clay minerals are layer silicates composed of sheets of alumina octahedra (an atom of aluminum bonded to six atoms of oxygen) or silica tetrahedra (an atom of silica bonded to four atoms of oxygen). Both tetrahedral (T) and octahedral (O) layers are organized around central Al,

Fe, or Mg cations. These sheets generally are bonded together in either a 1:1 structure (TO) in which each layer of alumina octahedra is paired with a layer of silica tetrahedra, or in a 2:1 structure (TOT) in which each octahedral layer is sandwiched between two tetrahedral layers. These building blocks are themselves interlayered and bound together by shared ions between the sheets. Layer architecture (1:1 vs. 2:1) and the ions between the sheets determine the physical properties of different clay minerals.

Adjacent layers in kaolinite, a clay mineral with a 1:1 structure, are held together by ionic bonds that are strong enough to prevent cations or water from entering the spaces between the sheets. Because it has few exchangeable cations held between its layers, kaolinite does not swell much when wetted, and has low plasticity (and thus little capacity to be molded). Clays with 2:1 layer structures exhibit much more variability in the chemical composition of their octahedral sheets (typically due to substitution of Fe^{2+} and Mg^{2+} for Al^{3+}) and in the abundance and type of ions present between layers. Smectite clays, like montmorillonite, have weak bonds between the silicate layers, which allows water and ions to readily penetrate the crystal structure. Also known as swelling clays, smectites expand readily upon wetting and are a main component of expansive soils. In expanding clays there are layers of water in the interlayer position, which explains why they can so readily take on or lose water and why they are so weak when expanded. Illite, the most common clay mineral in soils, has a strongly bonded 2:1 structure and fewer exchangeable cations between its layers, so it has less swelling potential than smectite.

Weathering of secondary minerals involves stripping off layers of silicate structure. The modification of muscovite (mica) to illite (clay), both of which consist of TOT "sandwiches" involves removal of the interlayer cations. Extreme weathering conditions can go beyond leaching of the intermediary cations and remove one of the two T layers, leaving a TO sequence of silicate layers, which is kaolinite clay. Stripping the remaining tetrahedral layer leaves the basic octahedral layer of gibbsite. In general, smectites weather to illites, and ultimately become kaolinite. Deeply weathered soils generally have high concentrations of kaolinite. It is worth noting, however, that kaolinite can form directly from primary minerals, depending on the parent material, climate, and intensity of weathering.

Process of Chelation

Chelation is the process through which relatively immobile metal ions, like iron and aluminum, are rendered mobile by soluble organic compounds that form ring structures around metal ions, making them susceptible to solution and transport. Chelation is particularly important in moving iron and aluminum, which are otherwise immobile in most soils. Chelation is facilitated by organic acids (particularly fulvic acid) produced by the breakdown of soil organic matter and by lichens, which produce chelating agents that accelerate weathering and liberate nutrients that help sustain the lichen. Iron and aluminum mobilized by chelating agents may be carried along in solution with soilwater flow until concentration changes or microbial actions break the chelating agent, causing the metal to reprecipitate. Incomplete conifer needle decay in cool, moist environments is a common source of chelating agents, leading to the stripping of Fe and other elements from forest soils.

Cation Exchange

An important outcome of chemical weathering is the ability of secondary minerals to exchange cations with soil water, thereby making nutrients available to plants. Clay minerals loosely hold exchangeable cations adsorbed on their surfaces. In many soils, the exchangable cations are associated mainly with organic matter. Ion exchange is the process by which ions in solution substitute for ions on mineral surfaces. Exchangeable cations are readily taken up in soil fluids, and they provide the dominant source of mineral nutrients for plants.

Clays and organic compounds vary in their ability to adsorb and release cations, a property called cation exchange capacity. Ion exchange is controlled by cation exchange capacity as well as by the ionic composition and pH of soil water.

Strongly acidic (low pH) pore fluids allow H+ to substitute for and replace metal cations. As hydrogen ions exchange places with nutrient cations held on a clay surface, the number of potentially exchangeable cations decreases.

The degree to which the exchange sites are occupied by exchangeable cations other than H^+ and Al^{3+} is called base saturation. Cation exchange progressively lowers base saturation in clay minerals by removing cations from between clay sheets. A clay with a high cation exchange capacity but a low base saturation has had its balancing cations stripped out and replaced by hydrogen ions through substantial chemical weathering. Such clays are typically found in tropical regions with high temperature and rainfall, such as Hawaii. The progressive loss of exchangeable cations reduces soil fertility — older, more intensively weathered soils are less fertile.

SOIL RESOURCES

The WRB Reference Soil Groups

After reviewing FAO's Revised Legend, 30 reference soil groups were identified to constitute the World Reference Base for Soil Resources. Three new reference soil groups are included: the *Cryosols, Durisols,* and *Umbrisols.* The *Greyzems* have been merged with the *Phaeozems,* and the *Podzoluvisols* are renamed *Albeluvisols.*

The 30 major soil groups of the WRB are *Acrisols, Albeluvisols, Alisols, Andosols, Anthrosols, Arenosols, Calcisols, Cambisols, Chernozems, Cryosols, Durisols, Ferralsols, Fluvisols, Gleysols, Gypsisols, Histosols, Kastanozems, Leptosols, Lixisols, Luvisols, Nitisols, Phaeozems, Planosols, Plinthosols, Podzols, Regosols, Solonchaks, Solonetz, Umbrisols,* and *Vertisols.*

Cryosols are introduced at the highest level to identify a group of soils which occur under the unique environmental conditions of alternating thawing and freezing. These soils have permafrost within 100 cm of the soil surface and are saturated with water during the period of thaw. In addition, they show evidence of cryoturbation. *Durisols* comprise the soils in semi-arid environments which have an accumulation of secondary silica, either in the form of nodules, or as a massive, indurated layer. *Umbrisols* cover the soils which have either an umbric horizon, or have a mollic horizon and a base saturation of less than 50 per cent in some parts within the upper 125 cm of the soil surface. They are a logical counterpart of the *Chernozems, Kastanozems* and *Phaeozems.*

The *Plinthosols* bring together the *Plinthosols* of the Revised Legend and the soils which have a petroplinthic layer at shallow depth. In the Revised Legend the latter soils belong to the *Leptosols.* For the World Reference Base it was decided to exclude from the *Leptosols* soils with pedogenetic horizons such as indurated calcic or gypsic horizons or hardened plinthite. This necessitated the definition of a reference soil group which included these soils. Although it is realised that soils with shallow petroplinthic layers and soils having plinthite normally occupy different positions in the landscape, it was felt appropriate to group them together as they are genetically related.

Podzoluvisols are renamed *Albeluvisols.* The name *Podzoluvisols* suggests that in these soils both the processes of cheluviation (leading to *Podzols*) and subsurface accumulation of clay (resulting in *Luvisols*) take place, while in fact the dominant process consists of removal of clay and iron/manganese along preferential zones (pea faces, cracks) in the argic horizon. The name *Albeluvisols* is therefore thought to be more appropriate, expressing the presence of a bleached eluvial horizon ("*albic horizon*"), a clay-enriched horizon ("*argic horizon*") and the occurrence of "*albeluvic tonguing*".

WRB Diagnostic Horizons, Properties and Materials

Earlier it was agreed that the soil groups should be defined in terms of a specific combination of soil horizons, called '*reference horizons*' rather than '*diagnostic horizons*'. Reference horizons were intended to reflect genetic horizons which are widely recognized as occurring in soils. Unfortunately, the distinction between reference and diagnostic horizons created confusion and it was agreed to retain the FAO terminology of diagnostic horizons as well as the diagnostic properties. Additionally it appeared necessary to define diagnostic soil materials. This together resulted in a comprehensive list of WRB diagnostic horizons, properties and materials, defined in terms of morphological characteristics and/or analytical criteria. In line with the WRB objectives, attributes are described as much as possible to help field identification.

Diagnostic Horizons and Properties

Of the 16 diagnostic horizons of the Revised Legend only *the fimic A horizon* has not been retained. It covers too wide a range of human-made surface layers and is replaced in the WRB by the hortic, plaggic and ferric horizons.

For the WRB, the definition of the *histic horizon* was broadened by reducing its minimum thickness to 10 cm and removing the maximum thickness. This is because of a second use of the definition. In the Revised Legend the histic H horizon is used to distinguish soils at second level to identify histic soil units; in the WRB it is used also at the highest level to define *Histosols*. It was agreed that *Histosols* over continuous hard rock should have a minimum thickness of 10 cm in order to avoid very thin organic layers over rock being classified as *Histosols*.

The P_2O_5 content requirement for FAO's mollic and umbric A horizons has been deleted from the WRB definition of *mollic* and *umbric horizons*. This requirement cannot be considered diagnostic since thick, dark coloured, human-made horizons in, for instance, China, also have low amounts of phosphate. Other criteria have to be found to separate mollic and umbric horizons from anthropedogenic horizons.

A *chernic horizon* is defined as a special kind of mollic horizon. The present definition of the mollic horizon was felt to be too broad to reflect properly the unique characteristics of the deep, blackish, porous surface horizons which are so typical for *Chernozems*.

The definition of the *ochric horizon* is similar to the ochric A horizon. The colour requirement for the *albic horizon* have been slightly changed compared to FAO's albic E horizon, to suit albic horizons which show a considerable shift in chrome upon moistening. Such conditions are frequently found in soils of the southern hemisphere.

The *argic horizon* definition differs from that of the argic B horizon of the Revised Legend in that the percentage clay skins on both horizontal and vertical ped faces and in pores has been increased from one to five per cent. This is expected to provide a better correlation with the earlier requirement of at least one per cent oriented clay in thin sections.

Guidelines to recognize a lithological discontinuity, if not clear from the field observation, were added to the description of the argic horizon. It can be identified by the percentage of coarse sand, fine sand and silt, calculated on a clay-free basis (international particle size distribution or using the additional groupings of the United States Department of Agriculture (USDA) system or other), or by changes in the content of gravel and coarser fractions. A relative change of at least 20 per cent in any of the major particle size fractions is regarded as diagnostic for a lithological discontinuity. However, it should only be taken into account if it is located in the section of the solum where the clay increase occurs and if there is evidence that the overlying layer was coarser textured.

The adjustments made in the description of the argic horizon also apply to the *natric horizon*. The definition of FAO's cambic B horizon has been slightly amended by deleting the requirement '*....and has at least eight per cent clay*'. This requirement forces some soils, which have a well-developed structural-B horizon and silt loam or silt textures with a low clay content, as found, for instance, in fluvio-glacial deposits of the nordic countries, into the *Regosols* rather than in the *Cambisols*. Because there is also no need for this requirement to separate *Cambisols* from*Arenosols* it has not been used in the definition proposed for the WRB *cambic horizon*.

Major alterations are made in the definition of the *spodic horizon*. It has been brought into line with the recent modifications in soil taxonomy (Soil Survey Staff, 1996) regarding the definition of spodic materials. Colour requirements were added, a limit of 0.5 or more in percentage oxalate extractable aluminium plus half that of iron is used, and a value for the optical density of oxalate extract (ODOE) of 0.25 or more is introduced. Moreover, the upper limit of spodic horizons has been set at 10 cm depth.

The silt-clay ratio of 0.2 or less has been deleted from the definition of the *ferralic horizon*. This criterion was felt to be too strict; the silt particle size fraction has been increased. Other values have been proposed (silt-clay ratio of 0.7 or less; fine silt-clay ratio of 0.2 or less) but, as yet, no consensus has been reached.

Some alterations are made in the definitions of the *calcic* and *gypsic horizons*. For WRB purposes they are split into calcic/gypsic and*hypercalcic / hypergypsic horizons*. These latter horizons have a calcium carbonate equivalent and gypsum content of 50 and 60 per cent, respectively, but are not cemented. The definition for the *sulfuric horizon* remains the same as in the Revised Legend. In addition to these diagnostic horizons, 19 new ones are proposed. Some are adopted from FAO's diagnostic properties, others are newly formulated. Together they bring the total of diagnostic horizons recognized in the WRB to 34. The newly defined diagnostic horizons are the *andic, anthropedogenic, chernic, cryic, duric, ferric, folic, fragic, fulvic, glacic, melanic, nitic, petroduric, petroplinthic, plinthic, salic, takyric, vertic, vitric* and *yermic horizons*.

A combination of an *anthraquic horizon* at the surface with an underlying *hydragric horizon*, totalling together a thickness of at least 50 cm, defines certain*Anthrosols* which show evidence of alteration through wet-cultivation practices. It comprises a puddled layer, a plough pan and an illuvial subsurface horizon. This combination is characteristic for soils which have been used for long-term paddy rice cultivation.

Newly defined diagnostic properties and materials are *albeluvic tonguing, alic* and *aridic properties*, and *anthropogeomorphic, calcaric, fluvic, gypsiric, organic, sulfidic* and *tephric soil material*.

Gleyic and *stagnic properties* have been reformulated. Slight changes are made in FAO's definitions of *abrupt textural change and geric properites*, while the definitions of *permafrost* and soft powdery lime, renamed *secondary carbonates*, have been adopted without change.

In the description of the *gleyic* and *stagnic properties* the occurrence of 'gleyic' and 'stagnic colour patterns' is introduced. These terms apply to the specific distribution pattern of Fe/Mn (hydr)oxides caused by saturation with groundwater or stagnating surface water. A gleyic colour pattern has 'oximorphic' features on the outside of structural elements, along root channels and pores, or as a gradient upwards in the soil. A stagnic colour pattern on the other hand shows these features in the centre of peas or as a gradient downwards resulting from impedance of the water flow.

The slight changes in the descriptions of *abrupt textural change* and *geric properties* refer to a different depth in which tile change in texture must occur and another way of calculating the effective cation exchange capacity (ECEC)[1], respectively.

EVOLUTION AND TOPOGRAPHY OF SOILS AND GEOMORPHOLOGY

Geomorphology (geo(Greek) = earth; morphos = form): the science that studies the evolution of topographic features by physical and chemical processes operating at or near the earth's surface. In geomorphology, the landscape is viewed as an assemblage of landforms which are individually transformed by landscape evolution. Because soils are an integral part of the landforms and landscape, processes occurring on the landscape have implications for soil processes. Conversely, soil processes can be considered to be a part of landscape evolution.

Landscape: the portion of the land surface that the eye can comprehend in a single view.

Landforms: distinctive geometric configurations of the earth's surface; features of the earth that together comprise the land surface

Geomorphic surface: a part of the surface of the land that has definite geographic boundaries and is formed by one or more agents during a given time span. It should be considered as a surface, *i.e.* similar to a plane, no thickness (z) - only x and y dimensions. Because it is formed during a specific time it is datable, either by absolute or relative means.

Erosion surface: a land surface shaped by the action of ice, wind, and water; a land surface shaped by the action of erosion

Constructional (depositional) surface: a land surface owing its character to the process of upbuilding, such as accumulation by deposition, for example fluvial, colluvial, or eolian deposits).

Geomorphic principles

Practically all work dealing with the distribution of soils on the earth's surface employs geomorphic concepts, and it is especially important in two areas;

1. Age, properties, and development rate of soils and
2. Hydrologic patterns on landscapes.

Soil Age

Soil development does not commence until erosion or deposition rate has reached a steady state that is less than the rate of soil formation. Thus, the age of the soil may or not be similar to the age of the deposits underlying the soil. For depositional surfaces, soil age is similar to the age of deposit (normally, slightly younger but related to the depositional event). Thus, radiocarbon or other dating methods of materials in the deposit are useful for determining soil age. This is not true for erosional surfaces. There may have been multiple erosion episodes since the material was deposited or exposed to sub-aerial processes. Thus, often the best that can be done is a relative age of the surface compared to other geomorphic surfaces in the area. Law of Superposition - younger beds occur on older beds if they have not been overturned.

SOILS AND LANDSCAPES

Soil properties and orders vary among climate zones and physiographic regions. Soil orders and the degree of soil development — particularly soil thickness and organic matter content — generally track latitudinal patterns in temperature and precipitation. Deeply weathered oxisols are typical of the equatorial tropics. Aridisols are typical of the mid-latitutde desert belts. Organic-rich mollisols and forest soils are typical of temperate latitudes. However, soils also differ within single landscapes because of variations in the factors of soil formation, erosion, and landscape history.

Soil Development

Soils mature over time. They gradually lose the physical and chemical characteristics of their parent materials, and take on characteristics that reflect soil-forming processes. Consequently, differences in parent material are better expressed in immature soils. More mature soils have characteristics determined by the dominant climate and vegetation.

Soil properties like the amount of organic matter, the degree and depth of oxidation, the removal of cations or minerals, development of clay minerals, and clay or $CaCO_3$ content in the B horizon change as the soil weathers and matures, so the degree of weathering can serve as a proxy for the stage of development and relative age of a soil.

The parts of a soil profile also mature at different rates. As a soil ages, the A horizon generally achieves maturity (*i.e.*, steady state or equilibrium) first, followed by the B horizon, and then such features as carbonate accumulation and oxidation. Different soil orders take different times to reach maturity too. Mature spodosols and mollisols are capable of forming in millennia, while mature ultisols may take tens of thousands of years to form and mature oxisols may require hundreds of thousands of years.

A chronosequence is a series of soils of different ages that formed on the same parent material under constant conditions of vegetation, topography, and climate, such as for example a sequence of soils formed on alluvial terraces at different heights above an incising river.

Chronosequences of soils that have been dated by radiometric or other means are used to evaluate rates of change in soil properties. Calibrated soil development rates can then be used to infer the ages of landforms from soil properties, or to estimate the amount of time that a soil was exposed at the surface before being buried.

Soil Catenas

Different kinds of soils can develop on different parts of a single slope because of local variability in soil forming processes, erosion, and deposition. A soil catena is a suite of soils with profiles that grade into one another in different landscape positions because of variations in soil moisture, sediment transport rates, slope steepness and chemical weathering. Catenas reflect local differences in soil-forming processes that result from topographic position and the hydrological and geomorphological processes that govern run-off, infiltration, and soil erosion.

For example, soils at the top of slopes where water readily washes over or through the soil profile may be well drained, oxidized, and reddish in colour, whereas those at the base of slopes tend to have a higher water table and reducing conditions that produce more gray to blue colours. Soils developed on the steep slopes ringing a plateau may be quite different from those developed on the flat plateau surface. In many environments, sloping surfaces are the rule and different soils develop at the top, mid-slope, and base of slopes. Soils in a catena reflect different records of the same processes acting on different parts of a landscape.

Paleosols

Most soils are geologically young, and actively evolving at the land surface. Paleosols, however, are ancient soils in which active soil processes have ceased. Some paleosols are buried soils preserved in the geologic record and some are relict soils that formed under different conditions but are still exposed at the surface. Buried soils are readily recognized, typically by an abrupt change in colour, and can be found in rocks of almost any geologic age. When found in the geologic record, paleosols indicate a land surface on which a soil was able to form.

They thus represent a period during which soil formation outpaced both erosion and burial by deposition. Because the characteristics of paleosols reflect the soil forming factors of the landscape in which they developed, they can be useful indicators of past climatic or environmental conditions. Some soil features, such as organic matter, do not survive extended burial and are poorly preserved in the rock record, whereas other characteristics, like soil structure and texture, are geologically robust.

Weathering-dominated Landforms

Weathering processes produce a variety of distinctive small- and large-scale landforms. At small scales, physical and chemical processes wear away the edges and corners of rock outcrops and boulders fastest, promoting spheroidal weathering that converts angular blocks of rock that result from tectonic and unloading fractures into rounded cobbles, boulders, and monoliths. Outward expansion of clay minerals on chemically altered rock surfaces enhances spheroidal weathering, and produces weathering rinds with an onion ring texture. Wetting and drying or freezing of water causes localized spalling and granular disintegration that creates weathering pits on bare rock surfaces of certain rock types. Extreme cases of weathering pit development result in cavernous or honeycomb textures known as tafoni. Dissolution pits and cavities often characterize highly soluble carbonate outcrops. At larger scales, weathering-dominated landforms include karst and thermokarst topography, inselbergs, tors, and duricrusts. Differential weathering in the valley and ridge province of the eastern United States results in quartzite ridgetops and limestone valley bottoms.

Inselbergs and Tors

Inselbergs (island mountains in German) and tors are high standing bodies of exposed rock that rise above surrounding terrain and result from spatial variability in weathering and erosion. Inselbergs are large residual rock masses still attatched to bedrock after episodes of deep weathering and erosion removed the surrounding material. Tors form the same way, but they are smaller features consisting of multiple, smaller exhumed corestones not necessarily still attached to bedrock.

What have today become inselbergs and tors weathered more slowly in the subsurface than surrounding areas because of differences in either mineral composition or fracture density that rendered them less susceptible to chemical weathering than the surrounding rock. Deep weathering produces an uneven weathering front, and inselbergs and tors represent exhumation of less weathered core stones protruding above an advancing weathering front. Australia's famous Uluru (also known as Ayers Rock), a massive sandstone outcrop that rises hundreds of meters above the desert plain of the central outback, is a prime example of a large inselberg.

Another is Rio de Janiero's famous Sugarloaf, a granitic inselberg towering above deeply weathered surroundings. Tors are common landforms in weathered landscapes around the world, but the rock towers of southwest England and Wales are well known examples. Inselbergs and tors record the changing balance between weathering and erosion or the change from an erosion-limited landscape to one where the pace of weathering limits erosion rates.

Duricrusts

Duricrusts are erosion-resistant cemented soils turned back into rock by cementation within the pedogenic zone. In arid regions, enough calcium carbonate ($CaCO_3$) may accumulate within a soil profile to form a cement-like layer of caliche, or calcrete, that is erosion resistant when it is exposed at the land surface. Silcrete, a hard silica-rich (SiO_2) layer, behaves similarly. Both calcrete and silcrete form in arid regions. Duricrusts also form by evaporation of solute-rich groundwater along stream valleys. These protective shells can even be tough enough to cause topographic inversions, wherein formerly low-lying terrain becomes more resistant to erosion than the neighbouring uplands and eventually becomes elevated as the surrounding terrain erodes away. Extreme oxidation, intense weathering and leaching produce iron-rich residual soils known as laterite and ferricrete. These erosion resistant crusts form where intense weathering removes all but the least mobile elements, leaving behind just aluminum and iron oxides. Ferricrete is an iron-rich duricrust made of highly concentrated iron oxides as a result of intensive weathering. Laterite and ferricrete are often found on slowly eroding topographic highs in the tropics, and form the caprock for mesas in central Australia and the Amazon. Evaporation of water within the soil or at the ground surface can coat mineral surfaces with precipitates. In arid and semi-arid regions like the Owens Valley in southeast California, soil water containing silica dissolved from fine dust and rock moves towards the surface and reprecipitates a hardened shell where it evaporates on the rock surface.

Applications

Weathering is of fundamental geomorphological importance because almost nothing happens on slopes without weathering to make slope-forming material transportable. The physical breakdown of rock into smaller pieces and chemical transformation into secondary minerals influence the types and rates of geomorphological processes that shape topography.

Soils define the frontier between geology and biology. The thin skin of weathered rock and decomposing organic matter provides the nutrients that nourish the plants on which all terrestrial life depends. Cation exchange capacity and base saturation of soils are the basis for soil fertility. Plants can readily extract nutrients from soils with high base saturation.

Sustained cultivation without replenishing soil nutrients leads to declining crop yields. This is one reason why floodplains that receive annual deposits of fresh minerals and volcanic soils that are periodically replenished by ash fall are prized and highly productive agricultural lands around the world. Modern industrial agriculture uses tremendous amounts of chemical fertilizers (principally nitrogen, phosphorus, and potassium) to supplement native soil fertility in place of traditional crop rotations, applications of manure, and organic farming techniques that are based on soil ecology. Understanding soil-forming processes and soil fertility are important for evaluating options for maintaining soil fertility and agricultural productivity in a post-petroleum (and potentially post-cheap fertilizer) world.

Maintenance of soil fertility depends not only upon maintaining soil nutrient levels, but upon conserving the soil itself. Extensive soil loss to erosion can follow deforestation, tillage, and destruction of vegetative cover by fire or overgrazing. On some now barren Caribbean islands, sugar cane cultivation on steep slopes sent most of the topsoil into the ocean within several generations. Recent earth history is rife with truncated and thinned soil profiles that record examples of ancient societies (such as Classical Greece, Rome, and Easter Island) that failed to prevent soil erosion from exceeding soil formation. Globally, the average rate of net soil loss from agricultural fields has been estimated to have increased by 10 to 20 fold as a result of tillage and exposure of bare soil to the effects of wind, rainfall, and run-off.

Farming practices have also reduced soil organic matter across vast areas of the continents, particularly in mollisols. About a third of the carbon dioxide added to the atmosphere by human activity since the Industrial Revolution came from the decay of soil organic matter resulting from plowing up fertile grassland soils, rather than from burning fossil fuels. But people can improve as well as degrade soil fertility.

The recently discovered organic-rich, incredibly fertile "terra preta" soils in the Amazon jungle formed over millennia as indigenous people burned their trash and broken pottery in their fields. Today these soils form islands of fertility in otherwise infertile tropical soils. Soils are among the most diverse features of Earth's surface, reflecting both regional environmental and local factors. Basic training in geomorphology can help tailor sustainable agricultural practices to particular soils and landforms.

Weathering also provides minerals critical for modern life. Deep weathering on ancient land surfaces produced iron and aluminum ores through pervasive leaching and removal of other common elements that concentrated relatively immobile elements in residual soils. Aluminum is a common element in terms of its distribution in Earth's crust, but it is dispersed at such low concentrations in most rocks that it is not economically extractable. It takes millions of years to dissolve away everything else and make laterite soils that are enriched enough in aluminum that they constitute aluminum ores. Few people realize that we warp our soda in ancient soils.

4

Geomorphology

Geomorphology is the study of landforms and landscapes, including the description, classification, origin, development, and history of planetar surfaces. During the early part of this century, the study of regional-scale geomorphology was termed "physiography". Unfortunately, physiography also became synonymouswith physical geography, and the concept became embroiled in controversy surrounding the appropriate concerns of that discipline. Some geomorphologists held to a geological basis for physiography and emphasized a concept of physiographic regions, A conflicting trend among geographers was to equate physiography with "pure morphology," divorced of its geological heritage.

In the period following World War II, the emergence of process, climatic, and quantitative studies led to a preference by many Earth scientists for the term "geomorphology" in order to suggest an analytical approach to landscapes rather than a descriptive one. In the second half of the twentieth century, the study of regional-scale geomorphology—the original physiography-was generally neglected. Russell attributed the decline of physiography to its elaborate terminology and to its detachment from evidence acquired by other disciplines, chiefly geology.

Although the concept of physiographic regions endured among geologists, geographers became much more interested in the details of man/land interactions and in the applications of modelling and systems analysis to geomorphology. In the exploration of planetary surfaces by various space missions, the perspective of regional geomorphology has been the required starting point for scientific enquiry. Global studies of Mars, the Moon, Mercury and Venus resulted in the identification of "surface units" or physiographic provinces. The Colorado Plateau is an excellent example of a terrestrial physiographic province. Plate I-1 illustrates the use of a large-scale perspective to focus on this naturally defined region. The term "mega-geomorphology" was introduced in March

1981 at the 21st anniversary meeting of the British Geomorpholgy Research Group. The proceedings of that meeting reveal that the concept was not well defined. It clearly involves a return by geomorphologists to the study of phenomena on large spatial scales, ranging from regions to continents to planets.

It also involves large time scales. Nevertheless, mega-geomorphology is merely a convenient term, unencumbered by past philosophical trappings, that emphasizes planetary surface studies at large scales. The interrelation of temporal and spatial scales in geomorphology is illustrated by the tentative classification. Of course, such a hierarchial ordering of geomorphic features is far from satisfying. As stated by Sparks, classifications are arbitrary constructions designed to facilitate the discussion of diverse phenomena at the risk of some distortion of the truth. The scheme merely illustrates what was well known to the great geomorphologists at the last turn of the century. The large first order features, continents and ocean basins, persist and evolve over long time scales. Small high-order features are transient. Fundamental units appear at different orders.

The old concept of physiographic regions was used to designate second-order forms, such as entire mountain ranges or coastal plains. Massive entities within a physiographic region might constitute a third-order form, such as a domal uplift. The details of the classification are unimportant as the analysis moves on to exploring the explanation of phenomena. This book explores mega-geomorphology. The parent science of geology has long emphasized large-scale features in its central discipline of tectonics. Although early proponents of largescale crustal mobilism, such as Alfred Wegener, were decidedly renounced by the mainstream scientific community, their ideas provided the stimulus for work that eventually transformed the Earth sciences. The plate tectonic model that emerged in the late 1960s was but a quantitatively geophysical confirmation of the elegant hypothesis developed by careful attention to large-scale structural patterns on the Earth´s surface.

Of course, this is not intended to imply that microscale studies are unimportant in structural geology. Such studies tell much about the details of rock deformation and the fabric of resulting materials. The session here is that significant science occurs at all scales of study. Scientists neglect the study of one spatial scale to the peril of their advancement to understanding.

Most geomorphologists would agree that certain fundamental assumptions underlie all geomorphological investigations. Whether termed "fundamental concepts", "philosophical assumptions", "paradigms", or "basic postulates", these ideas constitute a "conventional wisdom" for the science. One such fundamental concept involves the inherent complexity of landscapes. This concept has impeded the development of grand theories that survive the test of explaining numerous local features.

CONTEMPORARY GEOMORPHOLOGY

Today, the field of geomorphology encompasses a very wide range of different approaches and interests. Modern researchers aim to draw out quantitative "laws" that govern Earth surface processes, but equally, recognize the uniqueness of each landscape and environment in which these processes operate. Particularly important realizations in contemporary geomorphology include:

1) that not all landscapes can be considered as either "stable" or "perturbed", where this perturbed state is a temporary displacement away from some ideal target form. Instead, dynamic changes of the landscape are now seen as an essential part of their nature.
2) that many geomorphic systems are best understood in terms of the stochasticity of the processes occurring in them, that is, the probability distributions of event magnitudes and return times. This in turn has indicated the importance of chaotic determinism to landscapes, and that landscape properties are best considered statistically. The same processes in the same landscapes does not always lead to the same end results.

Processes

Modern geomorphology focuses on the quantitative analysis of interconnected processes. Modern advances in geochronology, in particular cosmogenic radionuclide dating, optically stimulated luminescence dating and low-temperature thermochronology have enabled us for the first time to measure the rates at which geomorphic processes occur. At the same time, the use of more precise physical measurement techniques, including differential GPS, remotely sensed digital terrain models and laser scanning techniques, have allowed quantification and study of these processes as they happen. Computer simulation and modelling may then be used to test our understanding of how these processes work together and through time.

Geomorphically relevant processes generally fall into (1) the production of regolith by weathering and erosion, the transport of that material, and its eventual deposition. Although there is a general movement of material from uplands to lowlands, erosion, transport, and deposition often occur in closely-spaced tandem all across the landscape.

The nature of the processes investigated by geomorphologists is strongly dependent on the landscape or landform under investigation and the time and length scales of interest. However, the following non-exhaustive list provides a flavour of the landscape elements associated with some of these.

Primary surface processes responsible for most topographic features include wind, waves, chemical dissolution, mass wasting, groundwater movement, surface water flow, glacial action, tectonism, and volcanism. Other more exotic geomorphic processes might include periglacial (freeze-thaw) processes, salt-mediated action, or extraterrestrial impact.

Fluvial Processes

Rivers and streams are not only conduits of water, but also of sediment. The water, as it flows over the channel bed, is able to mobilize sediment and transport it downstream, either as bed load, suspended load or dissolved load. The rate of sediment transport depends on the availability of sediment itself and on the river's discharge. Rivers are also capable of eroding into rock and creating new sediment, both from their own beds and also by coupling to the surrounding hillslopes. In this way, rivers are thought of as setting the base level for large scale landscape evolution in nonglacial environments. Rivers are key links in the connectivity of different landscape elements.

As rivers flow across the landscape, they generally increase in size, merging with other rivers. The network of rivers thus formed is a drainage system and is often dendritic, but may adopt other patterns depending on the regional topography and underlying geology.

Aeolian Processes

Aeolian processes pertain to the activity of the winds and more specifically, to the winds' ability to shape the surface of the Earth. Winds may erode, transport, and deposit materials, and are effective agents in regions with sparse vegetation and a large supply of unconsolidated sediments. Although water and mass flow tend to mobilize more material than wind in most environments, aeolian processes are important in arid environments such as deserts.

Hillslope Processes

Soil, regolith, and rock move downslope under the force of gravity via creep, slides, flows, topples, and falls. Such mass wasting occurs on both terrestrial and submarine slopes, and has been observed on Earth, Mars, Venus, Titan and Iapetus. Ongoing hillslope processes can change the topology of the hillslope surface, which in turn can change the rates of those processes. Hillslopes that steepen up to certain critical thresholds are capable of shedding extremely large volumes of material very quickly, making hillslope processes an extremely important element of landscapes in tectonically active areas.

On Earth, biological processes such as burrowing or tree throw may play important roles in setting the rates of some hillslope processes.

Glacial Processes

Glaciers, while geographically restricted, are effective agents of landscape change. The gradual movement of ice down a valley causes abrasion and plucking of the underlying rock. Abrasion produces fine sediment, termed glacial flour. The debris transported by the glacier, when the glacier recedes, is termed a moraine. Glacial erosion is responsible for U-shaped valleys, as opposed to the V-shaped valleys of fluvial origin.

The way glacial processes interact with other landscape elements, particularly hillslope and fluvial processes, is an important aspect of Plio-Pleistocene landscape evolution and its sedimentary record in many high mountain environments. Environments that have been relatively recently glaciated but are no longer may still show elevated landscape change rates compared to those that have never been glaciated. Nonglacial geomorphic processes which nevertheless have been conditioned by past glaciation are termed paraglacial processes. This concept contrasts with periglacial processes, which are directly driven by formation or melting of ice or frost.

Tectonic Processes

Tectonic effects on geomorphology can range from scales of millions of years to minutes or less. The effects of tectonics on landscape are heavily dependent on the nature of the underlying bedrock fabric that more less controls what kind of local morphology tectonics can shape. Earthquakes can, in terms of minutes, submerge large extensions creating new wetlands.

Isostatic rebound can account for significant changes over thousand or hundreds of years, and allows erosion of a mountain belt to promote further erosion as mass is removed from the chain and the belt uplifts. Long-term plate tectonic dynamics give rise to orogenic belts, large mountain chains with typical lifetimes of many tens of millions of years, which form focal points for high rates of fluvial and hillslope processes and thus long-term sediment production.

Features of deeper mantle dynamics such as plumes and delamination of the lower lithosphere have also been hypothesised to play important roles in the long term (> million year), large scale (thousands of km) evolution of the Earth's topography. Both can promote surface uplift through isostasy as hotter, less dense, mantle rocks displace cooler, denser, mantle rocks at depth in the Earth.

Igneous Processes

Both volcanic (eruptive) and plutonic (intrusive) igneous processes can have important impacts on geomorphology. The action of volcanoes tends to rejuvenize landscapes, covering the old land surface with lava and tephra, releasing pyroclastic material and forcing rivers through new paths. The cones built by eruptions also build substantial new topography, which can be acted upon by other surface processes.

Biological Processes

The interaction of living organisms with landforms, or biogeomorphologic processes, can be of many different forms, and is probably of profound importance for the terrestrial geomorphic system as a whole. Biology can influence very many geomorphic processes, ranging from biogeochemical processes controlling chemical weathering, to the influence of mechanical processes like burrowing and tree throw on soil development, to even controlling global erosion rates through modulation of climate through carbon dioxide balance. Terrestrial landscapes in which the role of biology in mediating surface processes can be definitively excluded are extremely rare, but may hold important information for understanding the geomorphology of other planets, such as Mars.

Overlap with other Fields

There is a considerable overlap between geomorphology and other fields. Deposition of material is extremely important in sedimentology. Weathering is the chemical and physical disruption of earth materials in place on exposure to atmospheric or near surface agents, and is typically studied by soil scientists and environmental chemists, but is an essential component of geomorphology because it is what provides the material that can be moved in the first place. Civil and environmental engineers are concerned with erosion and sediment transport, especially related to canals, slope stability (and natural hazards), water quality, coastal environmental management, transport of contaminants, and stream restoration. Glaciers can cause extensive erosion and deposition in a short period of time, making them extremely important entities in the high latitudes and meaning that they set the conditions in the headwaters of mountain-born streams; glaciology therefore is important in geomorphology.

CONTEMPORARY PERSPECTIVES AND GEOMORPHOLOGY

Geomorphologists will develop better understanding of the scientific foundations of their field through philosophical analysis. Although philosophical self-examination could be conducted in an independent manner, it is best pursued in the context of ongoing debate in the philosophy of the physical sciences. This type of analysis allows geomorphologists to address questions such as: what constitutes a scientifically valid explanation; are there different types of acceptable explanations, and, if so, what relationships, if any, exist among them; what is a geomorphological theory; to what extent is geomorphological knowledge a product of social, cultural, political, and ethical factors; are geomorphologists problem solvers or truth seekers; how do geomorphologists justify and defend their claims to knowledge; are these methods consistent with the aims of the field; and how are responses to these questions similar to or different from those provided by other scientific disciplines?

Although social constructivism, postpositivist empiricism, and scientific realism are the primary focuses of current philosophical debate, other viewpoints also may contain ideas relevant to geomorphology. Specifically, it may be worthwhile to explore philosophical perspectives that have emerged in disciplines such as biology and physics. However, there is no reason to presuppose that a philosophical framework for geomorphology will be merely a restatement of the philosophy of another discipline. Because geomorphology is concerned with distinctive types of natural systems that include synergistic physical and biological elements and employs characteristic investigative methods, it cannot be reduced to the underpinning disciplines.

One predominant idea in the philosophy of science merits serious consideration by geomorphologists. The three contemporary philosophical perspectives discussed above, though widely divergent in many respects, uniformly subscribe to the theory-laden view of scientific observation. Given their many differences, this commonality suggests something of uncommon significance. It also contrasts sharply with the theory-neutral interpretation of observation that appears to be a de facto—albeit often unconscious—belief of many geomorphologists. Philosophers have devoted considerable attention to the topic of scientific observation. Although a universally accepted definition has not emerged from this analysis, most philosophers agree that scientific observation cannot be equated only with perception; instead, it is a complex process that includes identification, interpretation, and description. The relationship between theory and observation in geomorphology deserves more consideration than it has received.

The philosophy of science not only provides a means of evaluating contemporary geomorphology but also serves as a framework for assessing the historical development of the discipline. Most available philosophical perspectives incorporate the explicit caveat that they apply to mature sciences only. Even scientific realists acknowledge that scientific theories may be largely social constructs in an inchoate science; therefore, it becomes difficult to assess whether a scientific discipline corresponds to the realist conception of scientific progress until it has developed a substantial history. Whether or not geomorphology qualifies as a mature science is unclear; nevertheless, any tentative assessment or debate about the nature of such an assessment should occur in the context of existing philosophical ideas.

In geomorphology the transition from the Davisian view of landscape development to the modern emphasis on systematic process-oriented investigations has been explicitly and implicitly characterised as a Kuhnian-style conceptual change or paradigm shift. On the assumption that the majority of geomorphologists prefer to view their field from a realist perspective, this characterisation of the historical development of geomorphology is inconsistent with the prevailing philosophical perspective, because the Kuhnian theory of science incorporates the notion that science does not progress towards the truth. The challenge for realists is to show how many theoretical constructs embodied in the Davisian view of geomorphology, including references to unobservables, have been preserved in contemporary geomorphic theories (Rhoads forthcoming). On the other hand, process geomorphologists must recognise that adopting a realist perspective does not necessitate that a truly scientific approach implies that all geomorphological problems must be described in the language of physics. Contemporary scientific realism explicitly acknowledges that no scientific discipline has privileged status with regard to the truth.

Exploration of philosophical issues in geomorphology should also enhance disciplinary unity. Contention in contemporary geomorphology centres on differences in regulative principles, types of scientific arguments, and characteristics of theory employed by scientists, all of whom consider themselves geomorphologists (Rhoads and Thorn 1993). Although the point of contention is scientific methodology, the contention itself is clearly philosophical in nature. In other words, it is not possible to resolve conflicts between competing methodologies within science itself; instead, resolution of these differences must occur within philosophy (Montgomery 1991). As this issue and others like it are examined philosophically, differences in opinion will emerge about which viewpoint provides the clearest perspective on geomorphology. However, here the goals of science and those of philosophy differ. Whereas practicing scientists define the success of their endeavours in empirical terms, such as confirmation and acceptance of factual knowledge, philosophers deal with concepts and define success quite differently.

Philosophers commonly exhibit great respect for the intellectual contributions of people who hold ideas directly opposed to their own. Success in philosophy is measured not by declaring that one perspective is right and that others are wrong, as tends to happen when methodological issues are debated in science, but by the degree to which opposing viewpoints have reached a consensus as to which issues are central to a specific debate and by the level of rational sophistication of the arguments. Philosophical insight by necessity depends on intellectual disagreement and debate. Viewed in this light, philosophical introspection provides an excellent antidote to scientists who wish to divide their colleagues into winners and losers on the basis of methodological preferences. Indeed, it offers the opportunity to embrace methodological diversity in a substantive and constructive manner and to enhance the intellectual depth of the discipline.

SCALES IN GEOMORPHOLOGY

Different geomorphological processes dominate at different spatial and temporal scales. Moreover, scales on which processes occur may determine the reactivity or otherwise of landscapes to changes in driving forces such as climate or tectonics. These ideas are key to the study of geomorphology today.

Overlap with other Fields

There is a considerable overlap between geomorphology and other fields. Deposition of material is extremely important in sedimentology. Weathering is the chemical and physical disruption of earth materials in place on exposure to atmospheric or near surface agents, and is typically studied by soil scientists and environmental chemists, but is an essential component of geomorphology because it is what provides the material that can be moved in the first place. Civil and environmental engineers are concerned with erosion and sediment transport, especially related to canals, slope stability (and natural hazards), water quality, coastal environmental management, transport of contaminants, and stream restoration. Glaciers can cause extensive erosion and deposition in a short period of time, making them extremely important entities in the high latitudes and meaning that they set the conditions in the headwaters of mountain-born streams; glaciology therefore is important in geomorphology.

TYPES OF GEOMORPHIC ANALYSIS

Regional landforms analysis can be approached with several different emphases. Since these derive from traditional geomorphic subdisciplines, this section will review several frameworks for study.

Process Studies and Systems Analysis

Process geomorphologists employ field, laboratory, and analytical techniques to study processes presently active on the landscape.

The work relies heavily on the incorporation of other disciplines, including pedology, soil mechanics, hydrology, geochemistry, remote sensing, hydraulics, statistics, geophysics, civil engineering, and geology.

To organize the complexities of process interactions, most geomorphologists utilize systems analysis. The landscape is idealized as a series of elements linked by flows of mass and energy.

Process studies measure the inputs, outputs, transfers, and transformations that characterize these systems. Although systems analysis does not constitute a true theory for geomorphology, it does serve the useful purpose of organizing process studies into a framework that allows modeling and prediction, especially when data are fed into digital computers. The systems approach to geomorphology has been extensively reviewed by Chorley and Kennedy and by Chorley et al.

Climatic Geomorphology

Climatic geomorphology developed as an alternative to Davisian theory for landscape evolution. Climatic geomorphologists hold that modern relief-forming mechanisms differ as a function of climate and that their relief products define major morphoclimatic zones on the globe.

Climatic geomorphologists systematized the various process combinations that occur in the morphoclimatic zones. Major practitioners include J. Tricart and A. Callieux of France and J. Budel of Germany.

German geomorphology is especially dominated by the approach of Budel. A related concept is climate-genetic geomorphology, which emphasizes the study of exogenic forces and especially climatic change as controls on the evolution of relief. Many climatic geomorphologists hold that little of the extant relief on the Earth is the product of modern relief-forming processes. Most is instead inherited from past morphoclimatic controls.

The study of relief generations or landscape evolution therefore consists of interpreting climatic changes in relation to certain diagnostic landscape features. An example of a useful designation of a morphoclimatic region is the periglacial zone. The term "periglacial" has come to mean the complex of cold-climate processes and landforms, including, but not limited to, those near active glaciers.

A key feature is frost action, especially the freezing and thawing of ground. A related, but not necessarily coincident, phenomenon is permafrost. Permafrost covers 20 to 25 per cent of the Earth's land surface.

It manifests itself on the landscape when large quantities of ground ice are present. This ice may form wedges that penetrate vertically into the regolith, growing with seasonal meltwater flow into tension cracks. Polygonal patterns characterize the ground surface.

Where ice-rich permafrost is degraded by geomorphic, vegetational, or climatic change, it forms a complex landscape known as thermokarst. Depressions form where zones of ground ice are removed by melting. In extreme cases, such as near Yakutsk in eastern Siberia, large valleys may form by the coalescence of thermokarst depressions.

PLANETARY GEOMORPHOLOGY

Some geomorphologists hold that their science is properly restricted either to the dynamic geology or to the physical geography of the Earth's surface. Indeed, more argument would probably be expended on the relative "geologic" or "geographic" content of geomorphology than whether any consideration should be given to bizarre alien landscapes. Such a view ignores two fundamentals. First, any science of the Earth must recognize that Earth is a planet. We learn more about that planet by studying analogs to its mysteries on other planets. Second, science derives its greatest excitement and its most important advancement through discovery. A century ago geomorphology was a science filled with wonder and excitement. The stimulus for its rapid growth in this time period was the discovery of landscapes that then seemed as bizarre and alien as those on other planets. The great plateaus of the western United States, the hyperarid deserts of the eastern understood terrestrial processes. Sharp emphasizes this benefit as follows:

"Planetary exploration has proved to be a two-way street. It not only created interest in Earth-surface processes and features as analogues, it also caused terrestrial geologists to look on Earth for features and relationships better displayed on other planetary surfaces." Because many planetary surfaces have been relatively stable for billions of years, they preserve the effects of extremely rare, exceedingly violent processes. Such processes include impact cratering, sturzstroms and cataclysmic flooding. On Earth, the evidence of such catastrophes is meagre because of rapid crustal recycling through plate tectonics and relatively high denudation rates. However, on the other terrestrial planets, the results of these processes can be studied in great detail, commonly in large-area images similar to those obtained by Earth-observing sensors in space.

This is of profound importance for Earth studies. When cataclysmic processes have occurred on Earth, their influence has been profound. The extinction of numerous organisms at the end of the Cretaceous because of a meteor impact is a case in point. The steppes of central Asia, the karst of Dalmatia, and the inselbergs of Australia and Africa all posed anomalies in the prevailing geomorphic theory.

The explanation of the new landscapes led to an expanded and improved explanation for landscapes already described. Of course, this reflects an obvious quality of all science: there are no bounds, geographic or otherwise, for enquiry into the origin of phenomena.

If a geomorphologist can learn more about the surface of Earth by studying other planetary surfaces, then that extraterrestrial study can no longer be dismissed as merely an interesting intellectual diversion. It becomes an absolutely essential part of geomorphology. An Earthcentered view of geomorphology is as limiting as a pre-Copernican view of the solar system. Not all geomorphologists have shared the modern reluctance to consider the study of extraterrestrial relief forms.

In 1892, the U.S. Geological Survey suffered a drastic cut in research funds. The Chief Geologist of the Survey at that time was geomorphologist Grove Karl Gilbert. Without support for his field work, Gilbert undertook a study using the U.S. Naval Observatory telescope in Washington, D.C., to compare surface features on the Moon with counterparts on Earth.

Despite the prevailing view that gradual and prolonged volcanism explained lunar surface features, Gilbert concluded that cataclysmic impact processes best explained the ubiquitous lunar craters. He applied the term "meteoric" to his impact theory, which had to wait over 70 years for verification by the Apollo programme of lunar landings and sample returns. Among the many sessions that geomorphology can trace to Gilbert's example that of studying other planets besides Earth-has yet to be fully appreciated. The study of planetary surfaces relies heavily on analogic reasoning to reconstruct the complex interactions of processes responsible for the observed landforms. Thus, the photointerpreter of planetary images must rely on this experience with terrestrial landscapes. Moreover, the geomorphic interpretation of other planets produces a kind of intellectual feedback: some planetary surfaces contain excellent analogs for little.

LANDFORMS ASSOCIATED WITH STREAMS

Floodplains

Most of the sediments on a floodplain are more or less continuously reworked as the stream channel meanders back and forth across the floodplain. Erosion occurs on the outside of the meanders, and deposition occurs on the point bar on the inside of the meanders. Point bar deposits are referred to as channel deposits in contrast to overbank deposits that settle out on the floodplain surface during floods.

Channel deposits are typically coarse grained (including gravel) while texture of overbank deposits vary across the floodplain, *i.e.* coarser deposits on natural levees adjacent to the channel, and clayey deposits in lower areas of the floodplain where floodwater ponds for a period of time allowing fine particles to settle. There are several other floodplain features that have characteristic particle size distribution and surface configuration. An important point is that both mineralogy and particle size of floodplain deposits depend on properties of the material in the source area from which it was eroded. If the headwaters of the stream are in an area with only silty soils and deposits, the floodplain deposits will contain little sand regardless of the landscape position within the floodplain at which they occur.

Stream Terraces

Channel incision responsible for formation of fluvial terraces results from a change in the flow regime of the stream.

Factors that may cause this change include:

- Change of base level, typically sea level change
- Tetonic uplift (changes elevation of upland in relation to base level (sea level)
- Climate change that affects load-discharge relationships. Change in rainfall amount or distribution may cause a change in erosion rates in the uplands and the amount of sediment delivered to the stream and/or a change in the flow velocity and transport capacity of the stream. Temperature changes may also affect evapotranspiration rate which will impact stream flow.

The simplest scenario is a sea level drop (lower base level) which increases stream gradient, flow velocity, and transport capacity. If sediment delivery to the stream remains constant, the added stream energy from the gradient increase may cause the channel bottom to erode and the stream to incise. When the stream reaches equilibrium with the new base level, it will again begin to meander, erode the valley walls (older alluvial deposits), and form a new floodplain at a lower elevation. The abandoned higher-elevation floodplain is referred to as a stream terrace.

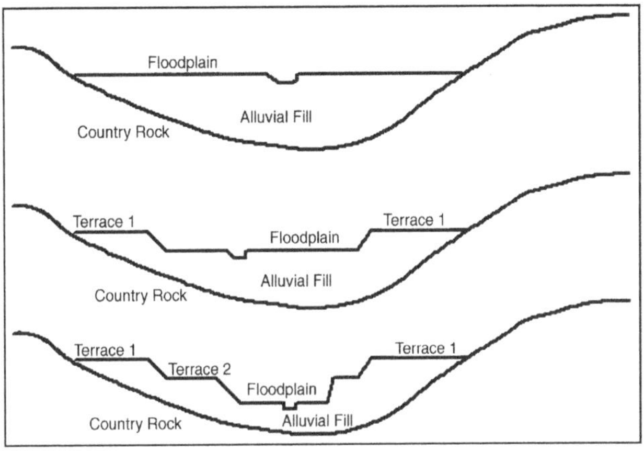

The diagram above illustrates paired terraces (same terrace level on each side of new floodplain) cut into older alluvium. There are many other forms of fluvial terraces. A major consideration associated with stream terraces is that the highest elevation terrace is the oldest and the terraces become progressively younger as elevation decreases. Degree of soil development will parallel terrace age and many of our concepts of time for development of pedogenic horizons and features have been developed from studies on fluvial terraces.

TECHNIQUES FOR MODERN LARGE-SCALE GEOMORPHIC ANALYSIS

It can be argued that the methodologies of a science are reflected in the research techniques of its practitioners. Indeed the recent review by Goudie clearly demonstrates the methodological focus of geomorphology on small-scale shortduration process studies. However, proper techniques do not guarantee proper results. Büdel provided a parable of a misguided process geomorphologist "of a generation who no longer read A. Penck."

This fictitious scientist studied processes on Alpine upland surfaces by "modern" methods, including soil analysis, grain-size distributions, clay mineralogy, slopewash monitoring, morphometry, and statistical analysis.

Büdel observes: "His conclusion was that these processes created the trough shoulders of the Alps.

His evidence for the certitude of these results was the indubitable precision of the analysis." The neglected fact was that the measured modern processes are all completely ineffective in modifying landforms that are relict from ancient times and that were formed by processes controlled by a completely different climate from that prevailing today.

Quantification

Quantification began to sweep geomorphology after the publication of R. E. Horton's visionary studies of drainage basin analysis. Some attempts at quantification were decidedly innovative. In the case of M. A. Melton, the work was so ahead of its time that only a few geomorphologists appreciated its implications. The extensive work on drainage basin and hillslope quantifications by Strahler and his students inspired a flowering of geomorphic research in the 1960s.

Similarly, the detailed studies of small-scale fluvial processes by Luna Leopold and colleagues at the U.S. Geological Survey also led to an abundance of related studies. Such process studies were decidedly advanced by technological developments that allowed for relatively easy measurement and long-term monitoring of processes in the field. Morphometric studies proved amenable to automatic data processing procedures by computer. Quantification has been described as a revolution in geomorphology. Although its use certainly superseded the qualitative approach of William Morris Davis, it is clear that quantification never constituted a revolution in the accepted sense of scientific philosophy. Quantification is a tool of study, one that indeed adds great power to the simplification of complexity. Nevertheless, it remains a mere technique, not a fundamental framework of thought.

One exciting aspect of quantification in mega-geomorphology derives from the ability of computing systems to handle the immense data sets necessary to describe terrain. The manipulation of these very large data sets will generate new and interesting research questions.

Role of Space Technology

Modern macrogeomorphology makes extensive use of global observations from spacecraft that employ a variety of imaging and sensing systems. These include vidicon imaging, multispectral scanning, radiometers, and radars. Modern image processing of digitally formatted data has revolutionized the interpretation of large-scale planetary landscape scenes.

The following question is posed: why have geomorphologists been slow to appreciate the global perspective afforded their planet by these advances? The question probably has many answers. The technology of remote sensing has only recently advanced to the point at which many geomorphologists can appreciate its relevance. The technology requires training in disciplines not normally considered in the training of geomorphologists. Even more interesting is the requirement placed on the user to have a large-scale view of problems.

Consider the perspective of a fluvial geomorphologist interested in floods. One approach might be to measure in excruciating detail the flood events he can easily access from his temperate climate university. This provides an impressive data set, but a problem remains. What of the immense rare floods that affect the great tropical regions of the planet? The immensity of flood effects in southwestern Queensland that followed phenomenal rains in 1974. Without satellite images, this scale of flooding could not have been measured at all.

Another solid example of the utility of appropriately processed space imagery is strikingly portrayed. This Landsat image of block mountains and basin fill in the Mojave Desert on the California and Arizona sides of the Colorado River between Lake Mohave and an agricultural development south of Parker. It has received both computer enhancement and special photographic reproduction. The resulting colour composite brings out both pronounced and subtle colour tones and patterns in the alluvial fill and pediments. In some parts of the scene, individual wash or fan patterns can be traced back to immediate source rock areas.

Three broad colour classes of unconsolidated surface materials can be recognized:

1. Light buff to tan material commonly in lower levels of the basin,
2. Reddishbrown material, and
3. Dark gray to blue gray material, forming aprons around many eroding ranges and generally overlapping the first class.

Although this false-colour rendition departs from the normal colours observed in the field or in natural colour aerial photography, it can be interpreted in terms of parent lithologies. Both the colour emphasis and the interactions among alluvial of deposits emanating from the mountain uplifts are far better displayed and interpreted synoptically than has been possible from aerial photographs, even after these are joined in mosaics.

In any science, new techniques are not important of themselves. It is rather the new discoveries made possible because of those techniques that stimulate scientific progress. A profound example of such a new discovery in terrestrial geomorphology came in November 1981 when the shuttle Columbia trained a space-age instrument on the Earth.

The Shuttle Imaging Radar carried by Columbia produced radar images of the hyperarid Selima Sand Sheet on the eastern Sahara.

The radar penetrated the sand cover to reveal fluvial valleys now filled by eolian sand. The valleys discovered by radar interpretation show a regional drainage system formed when the modern eolian-dominated landscape was subject to extensive fluvial erosion, probably during pluvial episodes of the Pleistocene and Tertiary.

5

Physical Geography and Earth's Interior

Physical geographers traditionally have avoided introspective philosophical analysis of their scholarly activities. This situation contrasts sharply with the fierce philosophical debate among human geographers during the past fifteen years. The primary purpose of this chapter is to persuade physical geographers of the virtue of engaging in philosophical introspection; however, it is important to recognise that physical geography is as diverse as human geography. This chapter focuses specifically on the potential contribution of philosophical analysis to geomorphology, but it also informs other sub fields of physical geography, because the issues discussed are central to all branches of the physical sciences. The reason for the neglect of philosophical issues by geomorphologists is unclear. Perhaps it reflects a basic sense of philosophical security. Although the issue of whether geomorphology should be regarded as scientific or historical has been somewhat controversial, much of this discussion has occurred within geology, not geography.

Contemporary geographical geomorphologists appear to be fairly secure in their role as physical scientists, a view that is reinforced by continued emphasis on the dichotomy between human and physical geography. Thus they may be confident that geomorphology is firmly rooted in established philosophical foundations of physics and chemistry. In other words, as physical scientists their philosophical underpinnings are incontrovertible, a perspective strengthened by the limited exposure that many geomorphologists have had to the philosophy of science.

Geomorphologists generally receive little, if any, formal training in philosophical topics at the undergraduate or graduate level. Moreover, in many cases what little training they do receive often comes from human geographers, who simply inform the physical geographers that they are empiricists and then proceed to survey the myriad philosophical perspectives in human geography. As a result most geographical geomorphologists have perfunctorily embraced either logical positivism (Harvey 1969) or critical rationalism (Haines-Young and Petch 1986). These philosophical perspectives generally are erroneously portrayed as equivalent, except for a minor difference in method: logical positivists attempt to verify hypotheses, but critical rationalists try to falsify them (Johnston 1991, 72).

One purpose of this chapter is to illustrate to human and physical geographers alike that the philosophy of the physical sciences, including geomorphology, is not as secure or uncontroversial as it may seem. Even if the philosophical foundation of the physical sciences is controversial, this situation may not trouble geomorphologists for two reasons.

First, they may presume that any debate about philosophical issues concerning the physical sciences should focus on physics or chemistry, not geomorphology, which is derivative from fundamental sciences. This view is misguided because geomorphology as a science deals with distinctive types of complex, unconstrained natural systems that differ from those investigated in laboratory sciences and in many cases uses methods that vary from those employed in laboratory investigations.

It is therefore uncertain that philosophical arguments developed for physics and chemistry directly inform geomorphology. Second, some geomorphologists probably perceive philosophy as an esoteric, unnecessary endeavour that has little relevance for practicing scientists.

Although occasional forays into philosophy are necessary for well-rounded scientists, to delve too deeply into such issues often is regarded with suspicion, if not outright contempt, and is sanctioned only if the primary purpose is to seek methodological guidance rather than philosophical insight. This disparaging attitude reflects a basic misunderstanding of the relationship between science and the philosophy of science.

Practicing scientists and philosophers of science have the same goal: to make sense of the world in which they live. Practicing scientists attempt to achieve the goal by interacting directly with the world. Philosophers of science contribute to attainment of the goal by examining the nature of scientific enterprise. Philosophers remind practitioners that they are part of the world, not neutral observers looking in from the outside. Philosophers provide a meta-enquiry into the strengths and weaknesses of science as a cognitive human enterprise directed towards attaining clear, certain, valid knowledge of the world.

This type of analysis provides the foundation for scientific claims to knowledge yet ensures that science as a whole does not become presumptuous about the certainty or truth of this knowledge. Thus any science has an essential relationship with philosophy. Because meaningful debate about the scientific foundation of geomorphology should at least consider, if not be grounded in, existing philosophical frameworks, the following discussion briefly surveys contemporary viewpoints in the philosophy of science, with a specific focus on perspectives pertinent to the physical sciences, and explores their relevance to geomorphology.

Contemporary viewpoints in the philosophy of science can be properly understood only when placed in the appropriate historical context. These viewpoints emerged mainly in response to growing dissatisfaction with logical positivism, or logical empiricism, which dominated the philosophy of science during the first half of the twentieth century (Suppe 1977). Logical positivism represented an attempt to demarcate science from non-science by eliminating metaphysical—that is, unobservable theoretical—elements from scientific theories.

The positivist view of science embraced the doctrines of knowledge empiricism, which maintains that all knowledge must be grounded in direct veridical phenomenal experience, and the verifiability theory of meaning, according to which the meaning of a scientific statement is determined by the test conditions under which it can be verified or disconfirmed.

Together these doctrines imply that scientific knowledge is restricted to knowledge of the observable. Knowledge of unobservable, theoretical entities is not possible: for the terms used to describe them to be meaningful, they must be explicitly or partially defined using verifiable, observational language. In other words, references to the unobservable have cognitive significance only if they are defined in terms of the observable. Thus the logical empiricist view of science is committed to metaphysical instrumentalism, or the notion that theoretical terms cannot refer to unobservable properties or entities and that sentences containing these terms do not have truth value (Matheson and Kline 1988).

Although logical empiricists do assign truth value to statements containing theoretical terms, this value is based not on a literal construal of these statements but on a rationally reconstructed construal that purges them of reference to the unobservable. Metaphysical instrumentalism implies that theories are merely logical calculating devices or instruments that allow scientists to derive observational predictions. Logical empiricists asserted that observation provides the foundation for knowledge and that empirical knowledge is accessible independent of theory, that is, theory neutral.

Theory development was viewed as an inductive process involving upward growth from observational facts to theoretical generalisations. Although logical positivists drew a clear distinction between the context of theory discovery and the context of theory justification, they did not directly address the issue of discovery. Instead they focused on the finished product of theorising. Scientific theories were perceived as hierarchical, mathematical constructs with high-level postulates connected deductively to low-level, testable observational statements via correspondence rules, that is, explicit or partial definitions.

Logical empiricists also maintained that scientific progress consists of the accumulation of verified empirical truths and a steady reduction in the number of scientific theories through amalgamation of competitors or absorption of small theories into comprehensive ones. The picture of science initially developed by logical positivists became known as the received view and evolved in a direction which attempted to avoid instrumentalism and to provide theories with a realist interpretation. Those efforts faced tremendous difficulties and rapidly degenerated as various elaborate and largely unsuccessful ad hoc modifications were implemented in an effort to salvage it (Suppe 1977). By the late 1960s many philosophers of science perceived the received view as fatally flawed.

The principal sources of dissatisfaction were its untenable distinction between the observable and the theoretical and its attempt to develop a normative philosophy describing the way in which scientists ought to conduct their business rather than accounting for contemporary and historical scientific activity and scientific claims to knowledge. Between 1960 and 1980 three alternative perspectives emerged as primary competitors in the contemporary philosophy of science: postpositivist empiricism, social constructivism, and scientific realism.

Although these three perspectives differ greatly, they share four commonalities: acknowledgment that the methods of science are theory dependent; increased emphasis on naturalistic philosophical perspectives that fully recognise the constraints imposed on theories of science by the judgements on which scientist base their claims to knowledge and by the natural history of science; elimination of the notion of theory as axiomatised, hierarchical, mathematical constructs and proliferation of less-formalised views of scientific theories; and recognition that no clear distinction exists between the context of discovery and the context of justification.

STRUCTURE OF THE EARTH'S INTERIOR

Studies with earthquake recordings have given a picture inside the Earth of a solid but layered and flow-patterned mantle about 2,900 km (1,800 miles) thick, which in places lies within 10 km (6 miles) of the surface under the oceans.

The thin surface rock layer surrounding the mantle is the crust, whose lower boundary is called the Mohorovièiæ discontinuity. In normal continental regions the crust is about 30 to 40 km thick; there is usually a superficial low-velocity sedimentary layer underlain by a zone in which seismic velocity increases with depth. Beneath this zone there is a layer in which *P*-wave velocities in some places fall from 6 to 5.6 km per second. The middle part of the crust is characterized by a heterogeneous zone with *P* velocities of nearly 6 to 6.3 km per second. The lowest layer of the crust (about 10 km thick) has significantly higher *P* velocities, ranging up to nearly 7 km per second.

In the deep ocean there is a sedimentary layer that is about 1 km thick. Underneath is the lower layer of the oceanic crust, which is about 4 km thick. This layer is inferred to consist of basalt that formed where extrusions of basaltic magma at oceanic ridges have been added to the upper part of lithospheric plates as they spread away from the ridge crests. This crustal layer cools as it moves away from the ridge crest, and its seismic velocities increase correspondingly.

Below the mantle lies a shell that is 2,255 km thick, which seismic waves show to have the properties of a liquid. At the very centre of the planet is a separate solid core with a radius of 1,216 km. Recent work with observed seismic waves has revealed three-dimensional structural details inside the Earth, especially in the crust and lithosphere, under the subduction zones, at the base of the mantle, and in the inner core. These regional variations are important in explaining the dynamic history of the planet.

Long-period Oscillations of The Globe

Sometimes earthquakes are large enough to cause the whole Earth to ring like a bell. The deepest tone of vibration of the planet is one with a period (the length of time between the arrival of successive crests in a wave train) of 54 minutes. Knowledge of these vibrations has come from a remarkable extension in the range of periods of ground movements that can be recorded by modern digital long-period seismographs that span the entire allowable spectrum of earthquake wave periods: from ordinary *P* waves with periods of tenths of seconds to vibrations with periods on the order of 12 and 24 hours such as those that occur in Earth tidal movements.

The measurements of vibrations of the whole Earth provide important information on the properties of the interior of the planet. It should be emphasized that these free vibrations are set up by the energy release of the earthquake source but continue for many hours and sometimes even days. For an elastic sphere such as the Earth, two types of vibrations are known to be possible.

In one type, called *S* modes, or spheroidal vibrations, the motions of the elements of the sphere have components along the radius as well as along the tangent. In the second type, which are designated as *T* modes, or torsional vibrations, there is shear but no radial displacements. The nomenclature is nSl and nTl, where the letters n and l are related to the surfaces in the vibration at which there is zero motion. Four examples are illustrated in the figure. The subscript n gives a count of the number of internal zero-motion (nodal) surfaces, and l indicates the number of surface nodal lines.

Several hundred types of *S* and *T* vibrations have been identified and the associated periods measured. The amplitudes of the ground motion in the vibrations have been determined for particular earthquakes, and, more important, the attenuation of each component vibration has been measured. The dimensionless measure of this decay constant is called the quality factor Q. The greater the value of Q, the less the wave or vibration damping. Typically, for ${}_0S_{10}$ and ${}_0T_{10}$, the Q values are about 250.

The rate of decay of the vibrations of the whole Earth with the passage of time, where they appear superimposed for 20 hours of the 12-hour tidal deformations of the Earth. At the bottom of the figure these vibrations have been split up into a series of peaks, each with a definite frequency, similar to that of the spectrum of light.

Such a spectrum indicates the relative amplitude of each harmonic present in the free oscillations. If the physical properties of the Earth's interior were known, all these individual peaks could be calculated directly. Instead, the internal structure must be estimated from the observed peaks.

Recent research has shown that observations of long-period oscillations of the Earth discriminate fairly finely between different Earth models. In applying the observations to improve the resolution and precision of such representations of the planet's internal structure, a considerable number of Earth models are set up, and all the periods of their free oscillations are computed and checked against the observations. Models can then be successively eliminated until only a small range remains. In practice, the work starts with existing models; efforts are made to amend them by sequential steps until full compatibility with the observations is achieved, within the uncertainties of the observations. Even so, the resulting computed Earth structure is not a unique solution to the problem.

SEISMIC AND THE EARTH'S STRUCTURE

The structure of Earth's deep interior cannot be studied directly. But geologists use seismic (earthquake) waves to determine the depths of layers of molten and semi-molten material within Earth. Because different types of earthquake waves behave differently when they encounter material in different states (for example, molten, semi-molten, solid), seismic stations established around Earth detect and record the strengths of the different types of waves and the directions from which they came. Geologists use these records to establish the structure of Earth's interior.

The two principal types of seismic waves are P-waves (pressure; goes through liquid and solid) and S-waves (shear or secondary; goes only through solid - not through liquid). The travel velocity of these two wave types is not the same (P-waves are faster than S-waves). Thus, if there is an earthquake somewhere, the first waves that arrive are P-waves. In essence, the gap in P-wave and S-wave arrival gives a first estimate of the distance to the earthquake.

As we know from physics, all waves change direction when they pass through layers of different density (refraction). That is what makes light collect in a magnifying glass, and that is also what makes seismic waves travel in curved paths through the Earth (because of the increasing pressure, materials are more dense towards the core, travel velocity of seismic waves increases). Refraction of seismic waves causes them to curve away from a direct path. Reflection causes them to glance off certain surfaces (e.g. core mantle boundary) when they hit it at too shallow of an angle. The result of this behavior, in combination with the fact that S-waves can not travel through liquids, is the appearance of seismic shadows, opposite of the actual earthquake site.

The geometric distribution and extent of these shadows as measured for a given earthquake (many receiver stations - seismographs, are needed all over the world to do that) allows us to calculate the position of major boundaries in the Earth's interior, as well as giving us information about the solid vs liquid character of the various layers, and even about some of their physical properties.

STRUCTURE OF THE EARTH

The interior structure of the Earth is layered in spherical shells, like an onion. These layers can be defined by either their chemical or their rheological properties. Earthhas an outer silicate solid crust, a highly viscous mantle, a liquid outer core that is much less viscous than the mantle, and a solid inner core. Scientific understanding of Earth's internal structure is based on observations of topography and bathymetry, observations of rock in outcrop, samples brought to the surface from greater depths by volcanic activity, analysis of the seismic waves that pass through Earth, measurements of the gravity field of Earth, and experiments with crystalline solids at pressures and temperatures characteristic of Earth's deep interior.

Assumptions

The force exerted by Earth's gravity can be used to calculate its mass, and by estimating the volume of the Earth, its average density can be calculated. Astronomers can also calculate Earth's mass from its orbit and effects on nearby planetary bodies.

Structure

The structure of Earth can be defined in two ways: by mechanical properties such as rheology, or chemically. Mechanically, it can be divided into lithosphere, asthenosphere, mesospheric mantle, outer core, and the inner core. The interior of Earth is divided into 5 important layers. Chemically, Earth can be divided into the crust, upper mantle, lower mantle, outer core, and inner core.

The layering of Earth has been inferred indirectly using the time of travel of refracted and reflected seismic waves created by earthquakes. The core does not allow shear waves to pass through it, while the speed of travel (seismic velocity) is different in other layers. The changes in seismic velocity between different layers causes refraction owing to Snell's law, like light bending as it passes through a prism. Likewise, reflections are caused by a large increase in seismic velocity and are similar to light reflecting from a mirror.

Core

The average density of Earth is 5,515 kg/m^3. Since the average density of surface material is only around 3,000 kg/m^3, we must conclude that denser materials exist within Earth's core. Seismic measurements show that the core is divided into two parts, a "solid" inner core with aradius of ~1,220 km and a liquid outer core extending beyond it to a radius of ~3,400 km. The densities are between 9.9 and 12.2 kg/m^3 in the outer core and 12.6–13.0 kg/m^3 in the inner core.

The inner core was discovered in 1936 by Inge Lehmann and is generally believed to be composed primarily of iron and some nickel. It is not necessarily a solid, but, because it is able to deflect seismic waves, it must behave as a solid in some fashion. Experimental evidence has at times been critical of crystal models of the core. Other experimental studies show a discrepancy under high pressure: diamond anvil (static) studies at core pressures yield melting temperatures that are approximately 2000K below those from shock laser (dynamic) studies. The laser studies create plasma, and the results are suggestive that constraining inner core conditions will depend on whether the inner core is a solid or is a plasma with the density of a solid. This is an area of active research.

In early stages of Earth's formation about four and a half billion (4.5×10^9) years ago, melting would have caused denser substances to sink toward the center in a process called planetary differentiation (see also the iron catastrophe), while less-dense materials would have migrated to the crust. The core is thus believed to largely be composed of iron (80%), along with nickel and one or more light elements, whereas other dense elements, such as lead and uranium, either are too rare to be significant or tend to bind to lighter elements and thus remain in the crust. Some have argued that the inner core may be in the form of a single iron crystal.

Under laboratory conditions a sample of iron nickel alloy was subjected to the corelike pressures by gripping it in a vise between 2 diamond tips, and then heating to approximately 4000 K. The sample was observed with x-rays, and strongly supported the theory that Earth's inner core was made of giant crystals running north to south.

The liquid outer core surrounds the inner core and is believed to be composed of iron mixed with nickel and trace amounts of lighter elements. Recent speculation suggests that the innermost part of the core is enriched in gold, platinum and other siderophile elements. The matter that comprises Earth is connected in fundamental ways to matter of certain chondrite meteorites, and to matter of outer portion of the Sun. There is good reason to believe that Earth is, in the main, like a chondrite meteorite. Beginning as early as 1940, scientists, including Francis Birch, built geophysics upon the premise that Earth is like ordinary chondrites, the most common type of meteorite observed impacting Earth, while totally ignoring another, albeit less abundant type, called enstatite chondrites. The principal difference between the two meteorite types is that enstatite chondrites formed under circumstances of extremely limited available oxygen, leading to certain normally oxyphile elements existing either partially or wholly in the alloy portion that corresponds to the core of Earth.

Dynamo theory suggests that convection in the outer core, combined with the Coriolis effect, gives rise to Earth's magnetic field. The solid inner core is too hot to hold a permanent magnetic field but probably acts to stabilize the magnetic field generated by the liquid outer core. The average magnetic field strength in Earth's outer core is estimated to be 25 Gauss, 50 times stronger than the magnetic field at the surface.

Recent evidence has suggested that the inner core of Earth may rotate slightly faster than the rest of the planet however more recent studies in 2011 found this hypothesis to be inconclusive. Options remain for the core which may be oscillatory in nature or a chaotic system. In August 2005 a team of geophysicists announced in the journal *Science* that, according to their estimates, Earth's inner core rotates approximately 0.3 to 0.5 degrees per year relative to the rotation of the surface.

The current scientific explanation for Earth's temperature gradient is a combination of heat left over from the planet's initial formation, decay of radioactive elements, and freezing of the inner core.

Mantle

Earth's mantle extends to a depth of 2,890 km, making it the thickest layer of Earth. Thepressure, at the bottom of the mantle, is ~140 GPa (1.4 Matm). The mantle is composed ofsilicate rocks that are rich in iron and magnesium relative to the overlying crust. Although solid, the high temperatures within the mantle cause the silicate material to be sufficiently ductile that it can flow on very long timescales. Convection of the mantle is expressed at the surface through the motions of tectonic plates. The melting point and viscosity of a substance depends on the pressure it is under. As there is intense and increasing pressure as one travels deeper into the mantle, the lower part of the mantle flows less easily than does the upper mantle (chemical changes within the mantle may also be important). The viscosity of the mantle ranges between 10^{21} and 10^{24} Pa·s, depending on depth. In comparison, the viscosity of water is approximately 10^{-3} Pa·s and that of pitch is 10^{7} Pa·s.

Crust

The crust ranges from 5–70 km (~3–44 miles) in depth and is the outermost layer. The thin parts are the oceanic crust, which underlie the ocean basins (5–10 km) and are composed of dense (mafic) iron magnesium silicate igneous rocks, like basalt. The thicker crust is continental crust, which is less dense and composed of (felsic) sodium potassium aluminium silicate rocks, like granite. The rocks of the crust fall into two major categories – sial and sima. It is estimated that sima starts about 11 km below the Conrad discontinuity (a second order discontinuity). The uppermost mantle together with the crust constitutes the lithosphere. The crust-mantle boundary occurs as two physically different events. First, there is a discontinuity in theseismic velocity, which is known as the Mohorovièiæ discontinuity or Moho. The cause of the Moho is thought to be a change in rock composition from rocks containing plagioclase feldspar (above) to rocks that contain no feldspars (below). Second, in oceanic crust, there is a chemical discontinuity betweenultramafic cumulates and tectonized harzburgites, which has been observed from deep parts of the oceanic crust that have been obducted onto the continental crust and preserved as ophiolite sequences. Many rocks now making up Earth's crust formed less than 100 million (1×10^8) years ago; however, the oldest known mineral grains are 4.4 billion (4.4×10^9) years old, indicating that Earth has had a solid crust for at least that long.

Historical Development of Alternative Conceptions

In 1692 Edmund Halley (in a paper printed in *Philosophical Transactions of Royal Society of London*) put forth the idea ofEarth consisting of a hollow shell about 500 miles thick, with two inner concentric shells around an innermost core, corresponding to the diameters of the planets Venus, Mars, and Mercury respectively. Halley's construct was a method of accounting for the (flawed) values of the relative density of Earth and the Moon that had been given by Sir Isaac Newton, in *Principia* (1687). "Sir Isaac Newton has demonstrated the Moon to be more solid than our Earth, as 9 to 5," Halley remarked; "why may we not then suppose four ninths of our globe to be cavity?"

EARTH'S GRID SYSTEMS

Topography, is the study of Earth's surface shape and features or those of planets, moons, and asteroids. It is also the description of such surface shapes and features. The topography of an area can also mean the surface shape and features themselves.

Planetary Energetic Grid Theory

Planetary Energetic Grid Theory falls under the heading of pseudoscience. It operates through geometric patterns called Sacred

Geometry. Grids meet at various intersecting points forming a grid or matrix. This is equivalent to the acupressure points on our bodies. These grid points can be found at some of the strongest power places on the planet. Plato recognized grids and their patterns, devising a theory that the Earth's basic structure evolved from a simple geometric shapes to more complex ones.

These shapes became known as platonic solids: cube, tetrahedron, octahedron, dodecahedron, icosahedron. In Timeaus, Plato associated each shape with one of the elements, earth, fire, air, ether, and water. The Earth's energy grids, from the beginnings of its evolutionary course, has evolved through each of these shapes to what it is today.

Each shape, superimposed, one upon the other to create a kind of all encompassing energy field that is the very basis of Earth holding it all together. Bill Becker and Bethe Hagens discussed the code of the Platonic Solids' positions on Earth, ascribing this discovery to the work of Ivan P. Sanderson, who was the first to make a case for the structure of the icosahedron at work in the Earth. He did this by locating what he referred to as Vile Vortices refer to a claim that there are twelve geometrically distributed geographic areas that are alleged to have the same mysterious qualities popularly associated with the Bermuda Triangle, the Devil's Sea near Japan, and the South Atlantic Anomaly.

Becker and Hagens' attention was drawn to this research through the work of Chris Bird, who punished "Planetary Grid" in the New Age Journal in May 1975. After meeting with Bird, they completed their Grid making it compatible with all the Platonic Solids, by inserting a creation from Buckminster Fuller's work. They proposed that the planetary grid map outlined by the Russian team Goncharov, Morozov and Makarov is essentially correct, with its overall organization anchored to the north and south axial poles and the Great Pyramid at Gizeh. They believed the Russian map lacked completeness, which led them to them overlaying a complex, icosahedrally-derived, spherical polyhedron developed by R. Buckminster Fuller.

In Synergetics, he called it the "Composite of Primary and Secondary Icosahedron Great Circle Sets." South America's grid triangle forms the continent around itself. In looking at the southernmost tip of South America, you can see how the force of node number 58 pushes the land away from its due south trend and towards the east. Then, if you look at node 49, on the middle of the East Coast of South America near Rio de Janeiro, you can clearly see how the force of the node has pushed the landmass of the continent into a rounded shape.

Looking at Australia, you can clearly see that the whole continent, and especially the northwest side, forms very precisely within the stretching forces of this area of the Global Grid. Here, if we look to the exact north and middle of Australia on point 27, we see a circular "node point" displacing the land around it and forming the Gulf of Carpenteria.

Again, the nodes themselves have shaped the land into circular "vortices," repelling the continental mass from themselves and in this case, also shaping the outline of the island directly above Australia. Node 44 is precisely aligned with the bottom of Antarctica, and we see either edge "drooping" to the right or the left from this. Richard Lefors Clark, this is the "bowtie" energetic configuration showing itself in the Australian landmass, which he calls a "diamagnetic energy vortex." As suggested, examine how such a shape could be formed by the curved, spiraling energies that make up the grid below. Clark also shows us that the Gulf of Mexico follows this same curving energetic organization, as well as the two coastlines of the continental United States.

Africa shows the combined action of two larger triangles, one with the point facing downwards on the African continent and its neighbour pointing upwards, griding the Indian Sea. The downward-pointing triangle certainly describes Africa well, and the extra land on the western side can be explained by the pressure coming from the Mid-Atlantic Ridge, which is almost precisely the same as the Atlantic grid line formed by points 10, 19, 37, 38, 39 and 50. The east coast of Africa and Saudi Arabia follow the eastern side of the triangle very nicely, on the grid line from point 41 to point 12. Node point 22 gently pushes in on Africa's east coast, causing it to dip inwards as it travels and forming the Somalia Basin.

We can also assume that the "cracks" separating Africa and Saudi Arabia were caused by the expansion of the Grid, forcefully pulling the land into alignment. The "crack" to the immediate left of point 12 follows the grid line exactly. And finally, the upward-pointing triangle can be clearly seen to cut its way right into the land, with Africa's coast defining its left side and India's coast defining its right. Node point 22 gently pushes in on Africa's east coast, causing it to dip inwards as it travels and forming the Somalia Basin.

We can also assume that the "cracks" separating Africa and Saudi Arabia were caused by the expansion of the Grid, forcefully pulling the land into alignment. The "crack" to the immediate left of point 12 follows the grid line exactly. And finally, the upward-pointing triangle can be clearly seen to cut its way right into the land, with Africa's coast defining its left side and India's coast defining its right. In this next illustration, we can see how point 21, in the centre of the African continent, works with its above-right partner point 20 in providing a framework for the bow-shaped energy vortex that shaped the northeast coast of Africa.

Point 20 is the only "vile vortex" that is significantly inland; 41, near South Africa, and 42, near India, are the only others that touch land at all! This is probably due to the incredible strength that they possess, which seems to repel landmasses. We can see how the northeast coast of Africa is indeed very circular, and point 20 is directly at the centre of this "bowtie" of energy. We see Dr. Lefors Clark's bow-shaped "diamagnetic energy vortex" shaping the land in a smooth curve. We can also see the same curving formation created by the east coast of India and the west coast of the China/Korea/Vietnam area of Asia.

This vortex is balanced between points 24 and 13. We remind ourselves that these smooth curves show the spiraling nature of the superstrings that make up these geometric energy fields, here expressed as spiraling lines of magnetic force. Dr. Clark, the point at the centre of the "bowtie" becomes a magnetic null zone of zero gravity when appropriately triggered by the right geometric positions of the Sun, Moon and Planets to the Earth.

In those moments of conjunction, there is a "hyperdimensional bleed-through," and we then get time dilations. The circular grid energy formations surrounding the east coast of China and surrounding area: We have our straight grid map of China for comparison. Three circular vortexes in it. We have already discussed the bottom left vortex in India and China.

Then, the centre vortex shapes the East Coast of China, based off of and emanating from point 13. The uppermost and largest vortex shapes the entire Japanese and Russian archipelago, and is centred in and emanating from point 4. We have a vortex centred in Sanderson's "vile vortex" near Japan, the "Devil's Sea," here listed as point 14. This vortex is "equalized" by points 25 and 26, both of which are equidistant from the edge of the circle. And obviously, Indonesia precisely traces the bottom left of the circle itself. This is obviously a very strong vortex to be able to shape the land of Indonesia from where it stands, just as to Becker and Hagens. Another area of continuing disappearances and mysterious time-warps is the Devil's Sea located east of Japan between Iwo Jima and Marcus Island. Here events have become so sinister that the Japanese government has officially designated the area a danger zone. This area was significant enough to the work of Bermuda Triangle author Charles Berlitz that he wrote a whole book dedicated to it and the "bigger picture," entitled The Devil's Triangle. It is becoming more and more clear that our geometric shapes, expressed as the consciousness unit expanded to a planetary scale, are far more than abstract concepts of theoretical physics. What we have here are direct, quantifiable and measurable phenomena, and these geometry-based Grid patterns are simply the simplest, and therefore the best solution to the problem.

The "grid bands" on the Earth and how their effects shaped the Ring of Fire', the flowing of the Nile River, the "node point" of the Egyptian northern coast centred in 'Giza, and the vertical structure of the Yucatan Peninsula. Now, with the full map of the Becker-Hagens grid, we can see a great deal more from the effect of these straight longitudinal lines. By referring back to the main map, the reader can clearly see that the entire Eastern side of Hudson Bay in Canada precisely follows the line from 18 in Florida to 9 in the Bay to 61 at the North Pole. Furthermore, all of England is precisely within the line created by point 20 in Africa, through point 11 in England to point 61 at the North Pole.

So, there are a variety of ways to see this energy at work on Earth. One can begin to visualize this Grid energy as a living net of "wires" that are tightly stretched over a thin balloon. It is obvious to us that what we think of as randomly placed continents are actually conforming to this massive energy, disappearances, gravity loss, levitation and other related phenomena. Becker-Hagens explain how these grid points seem to attract large population centres. Look at the South American landmass. Not only does it fit perfectly a Grid Triangle, but we can see a circular, bowtie-shaped energy at work in the actual shaping of the landmass itself. This "bowtie" is nearly centred within the diamond that is formed between points 18, 35, 37 and 49. The offset of the South American landmass from being precisely within this "diamond" again could be accounted for by the pushing of the Mid-Atlantic Ridge, which follows the lines of the Global Grid with amazing precision.

Looking back at the Becker-Hagens map, it is quite easy for us to see how this line clearly demarcates the separation between the continents, just as the Mid-Atlantic Ridge is the point of expansion between the two continental plates. An elliptical-shaped gravity field is clearly visible when the centre is placed directly in node 15. If we use any standard image editing programme and "drag out" an ellipse using point 15 as a centre, we can indeed align it precisely with the island formations to the far west of our diagram.

We can see the amazing connections of this energy formation as soon as we start looking at the diagram. We can try other centre points besides 15, but the ellipse will not fit anywhere else as precisely as it does right at that spot. This massive energy vortex seems to provide the clearest Grid counterpart for the existence of the Ring of Fire, which is a ring of volcanoes and tectonic activity surrounding the Pacific Ocean. When we look at this "grid ring" carefully, we can see that it represents the perfect fusion between the Earth's landmasses and the Global Grid.

Going clockwise from the 12:00 point, the ring will perfectly touch a "square" of grid points. We can also see that this ellipse is well defined by points 14 and 16, again Sanderson's incredibly powerful "vile vortices," the points of the icosahedron.

We have already seen how the incredible gravitational force of these "vortices" was able to shape the entire upper Western half of Africa into an elegant, circular shape. Now, we are seeing two of these vortices working together to form an even larger shape.

The ring forms part of the East Coast of China, as well as a good part of the upper Russian coastline surrounding node 5. It also defines part of the southern coastline of Alaska. The grid points 14 and 16 would be akin to the two poles of the dividing cell. The grid lines provide us with a simplified depiction of the "spindle fibres" that form in cell mitosis. The actual ring of energy that is created forms a precise analog of the nuclear membrane of the cell, as it continues its expanding, elliptical process of division.

Ley Lines

Ley lines are alleged alignments of a number of places of geographical interest, such as ancient monuments and megaliths that are thought by certain adherents to dowsing and New Age beliefs to have spiritual power. Their existence was suggested in 1921 by the amateur archaeologist Alfred Watkins, *The Old Straight Track.*

The believers in ley lines think that the lines and their intersection points resonate a special psychic or mystical energy. Ascribing such characteristics to ley lines has led to the term being classified as pseudoscience. Ley lines can be the product of ancient surveying, property markings, or commonly traveled pathways. Many cultures use straight lines across the landscape. In South America, such lines often are directed towards mountain peaks; the Nazca lines are a famous example of lengthy lines made by ancient cultures.

Straight lines connect ancient pyramids in Mexico; today, modern roads built on the ancient roads deviate around the huge pyramids. The Chaco culture of Northwestern New Mexico cut stairs into sandstone cliffs to facilitate keeping roads straight. Additionally, chance alignments and coincidence are often cited as explanations that cannot be ruled out. The concept of ley lines was first proposed by Alfred Watkins.

On June 30, 1921 after Watkins visited Blackwardine in Herefordshire, and went riding a horse near some hills in the vicinity of Bredwardine. There he noted that many of the footpaths there seemed to connect one hilltop to another in a straight line. He was studying a map when he noticed places in alignment. "The whole thing came to me in a flash", he later told his son.

However, in September 1870, William Henry Black gave to the British Archaeological Association, in Hereford, a talk titled Boundaries and Landmarks, in which he speculated that "Monuments exist marking grand geometrical lines which cover the whole of Western Europe". It is possible that Watkins's experience stemmed from faint memories of an account of that presentation.

Watkins believed that, in ancient times, when Britain was far more densely forested, the country was crisscrossed by a network of straight-line travel routes, with prominent features of the landscape being used as navigation points. This observation was made public at a meeting of the Woolhope Naturalists' Field Club of Hereford in September 1921. His work referred to G. H. Piper's paper presented to the Woolhope Club in 1882, which noted that: "A line drawn from the Skirrid-fawr mountain northwards to Arthur's Stone would pass over the camp and southern most point of Hatterall Hill, Oldcastle, Longtown Castle, and Urishay and Snodhill castles."

The ancient surveyors who supposedly made the lines were given the name "dodmen". Watkins published his ideas in the books Early British Trackways and The Old Straight Track. They generally met with skepticism from archaeologists, one of whom, O. G. S. Crawford, refused to accept advertisements for the latter book in the journal Antiquity. Most archaeologists since then have continued to reject Watkins's ideas. Despite the mostly negative reception to his ideas, some experts have made observations similar to Watkins's.

Megalithic researcher Alexander Thom offered a detailed analysis of megalithic alignments, proposing a standardization of measure by those who built megaliths, but avoided the term ley line. The discovery by Europeans of the Nazca lines, man-made lines on desert pavement in southern Peru, prompted study of their astronomical alignments. The existence of alignments between sites is easily demonstrated. However, the causes of these alignments are disputed.

There are several major areas of interpretation:

- *Archaeological*: A new area of archaeological study, archaeogeodesy, examines geodesy as practiced in prehistoric time, and as evidenced by archaeological remains. One major aspect of modern geodesy is surveying. As interpreted by geodesy, the so-called ley lines can be the product of ancient surveying, property markings, or commonly travelled pathways. Numerous societies, ancient and modern, employ straight lines between points of use; archaeologists have documented these traditions. Modern surveying also results in placement of constructs in lines on the landscape. It is reasonable to expect human constructs and activity areas to reflect human use of lines.
- *Cultural*: Many cultures use straight lines across the landscape. In South America, such lines often are directed towards mountain peaks; the Nazca lines are a famous example of lengthy lines made by ancient cultures. Straight lines connect ancient pyramids in Mexico; today, modern roads built on the ancient roads deviate around the massive pyramids. The Chaco culture of Northeastern New Mexico cut stairs into sandstone cliffs to facilitate keeping roads straight.
- *New Age*: Some writers widely regarded as pseudoscientific have claimed that the ley lines and their intersection points resonate a special psychic or magical energy. These theories often include elements such as geomancy, dowsing or UFOs. Some similar believe these points on lines have electrical or magnetic forces associated with them.
- *Skeptical*: Skeptics of the actuality of ley lines often classify them as pseudoscience. Such skeptics tend to doubt that ley lines were planned or made by ancient cultures, and argue that apparent ley lines can be readily explained without resorting to extraordinary or pseudoscientific ideas.

Spiritual Significance of Ley Lines: Magical and Holy Lines

Watkins's theories have been adapted by later writers. Some of his ideas were taken up by the occultist Dion Fortune who featured them in her 1936 novel The Goat-footed God. Since then, ley lines have become the subject of a few magical and mystical theories. Two British dowsers, Captain Robert Boothby and Reginald A. Smith of the British Museum, have linked the appearance of ley lines with underground streams and magnetic currents. Guy Underwood conducted various investigations and claimed that crossings of 'negative' water lines and positive aquastats explain why certain sites were chosen as holy. He found so many of these 'double lines' on sacred sites that he named them 'holy lines.' Separate from other spiritual theories of ley lines, two German Nazi researchers Wilhelm Teudt and Josef Heinsch have claimed that ancient Teutonic peoples contributed to the construction of a network of astronomical lines, called [3]Holy lines[2], which could be mapped onto the geographical layout of ancient or sacred sites.

Teudt located the Teutoburger Wald district in Lower Saxony, centred around the dramatic rock formation called Die Externsteine as the centre of Germany. Nazism often employed ideation of superiority and associated Aryan descent with ancient higher cultures, often without regard for archaeological or historic fact.

Chance Alignments

Watkins's discovery happened at a time when Ordnance Survey maps were being marketed for the leisure market, making them reasonably easy and cheap to obtain; this may have been a contributing factor to the popularity of ley line theories. Given the high density of historic and prehistoric sites in Britain and other parts of Europe, finding straight lines that "connect" sites is trivial, and ascribable to coincidence. The diagram to the right shows an example of lines that pass very near to a set of random points: for all practical purposes, they can be regarded as nearly "exact" alignments.

For a mathematical treatment of this topic. Since the existence of alignments themselves are not controversial, analysis can proceed by an attempted rejection of the null hypothesis that ley-line-like alignments are due to random chance. Statistical analysis by skeptics of this hypothesis shows that random chance is consistent with the evidence.

Some Chaos Magicians claim such results to be in accord with their generative view of chance, though such alternative null hypothesis explanations are usually deprecated on philosophical grounds in hypothesis testing due to considerations of falsifiability and Occam's razor.

One study by David George Kendall used the techniques of shape analysis to examine the triangles formed by standing stones to deduce if these were often arranged in straight lines. The shape of a triangle can be represented as a point on the sphere, and the distribution of all shapes can be thought of as a distribution over the sphere.

The sample distribution from the standing stones was compared with the theoretical distribution to show that the occurrence of straight lines was no more than average. Archaeologist Richard Atkinson once demonstrated this by taking the positions of telephone boxes and pointing out the existence of "telephone box leys".

This, he thus argued, showed that the mere existence of such lines in a set of points does not prove that the lines are deliberate artifacts, especially since it is known that telephone boxes were not laid out in any such manner, and without any such intention.

Straight lines also do not make ideal roads in many circumstances, particularly where they ignore topography and require users to march up and down hills or mountains, or to cross rivers at points where there is no portage or bridge.

Examples

Alfred Watkins identified St. Ann's Well in Worcestershire as what he believed to be the start of a ley line that passes along the ridge of the Malvern Hills through several springs including the Holy Well, Walms Well and St. Pewtress Well. In the late 1970's Paul Devereux stated he had discovered the Malvern Ley which began at St Ann's Well and ended at Whiteleaved Oak. The alignment passes through St. Ann's Well, the Wyche Cutting, a part of the Shire Ditch, Midsummer hillfort and Whiteleaved Oak.

Hartmann Net or Hartmann Lines

The Hartmann net consists of naturally occurring charged lines, running North-South and East-West. It is named after Dr. Ernst Hartmann, a well regarded German medical doctor, who first described it soon after the second world war. Alternate lines are usually positively and negatively charged, so where the lines intersect it is possible to have double positive charges and double negative charges, or one positive and one negative charge. It is the intersections that are seen to be a source of potential problems. The Hartmann Net appears as a structure of radiations rising vertically from the ground like invisible, radioactive walls, each 21 centimetres wide. The grid is magnetically orientated, from North to South they are encountered at intervals of 2 metres, while from East to West they are 2.5 metres apart. Between these geometric lines lies a neutral zone, an unperturbed micro-climate. This network penetrates everywhere, whether over open ground or through dwellings.

The Hartmann net has been defined using the Chinese terms of Yin and Yang. The Yin is a cold energy which acts slowly, corresponds to winter, is related to cramps, humidity and all forms of rheumatism. The

Yang is a hot, dry rapidly acting energy. It is related to fire and is linked to inflammations. The points formed by the intersection of these lines, whether positive or negative, are dynamic environments sensitive to the rhythms of the hours and the seasons. It has been suggested that both the Curry grids and Hartmann Net are earthing grids for cosmic rays that constantly bombard the Earth, and that they can be distorted by other things, such as geological fault lines and underground mining. It is also possible to have spots where the Curry and Hartmann lines cross, causing further potential problems. These spots are generally seen to be more detrimental than a single crossing within the Curry or Hartmann system.

Curry Lines

Earth radiation is a hypothetical geophysical phenomenon described primarily by the German authors Manfred Curry and Ernst Hartmann. This is known as Curry Lines. Both men describe a mystic force field, similar to Odic force, Mana, and Qi, that covers the Earth at regular intervals and can be detected by dowsing using a divining rod.

It is not supposed to be detectable by common scientific instruments but some still connect it to telluric currents, which are actual phenomena, detectable by scientific instruments. Placing people or other living things in certain spots of the earth radiation knots is believed to be beneficial/ harmful depending on radiation flow direction.

It connects to the Gaia philosophy and vitalist school and is very popular in certain New Age circles in Europe, especially in Germany. The radiation is described as a grid-like arrangement with lines at regular distances

Comparing Curry Lines, Hartmann Lines and Ley Lines

Curry lines are approximately 3 meters apart, diagonally to the poles, east to west. Hartmann lines run both east-west and north-south forming a grid across the earths surface with a distance of circa 2 meters in the north-south direction and 2.5 meters in the east-west direction. Ley lines are man-made energy lines, created by stone formations such as stone ships or other ancient archaeological structures. The knowledge of creating Ley lines is supposed to be lost.

Black Lines

Black lines seem to be naturally generated, although quite how is not known. They may be localized and do not form a network in the same way as Hartmann and Curry lines. This could be similar in nature to the "sha", or deadly energy lines of Chinese Feng-Shui. They can be curved, straight, at ground level or higher, even found in the upper levels of buildings. There have been described 2 types of Black lines, one as "black and depressed", the other as "shiny, black, hard and sharp." They could possibly represent the flow lines of a negative type of "orgone-type" energy as described by Wilhelm Reich.

THE EARTH'S INTERNAL HEAT ENERGY AND INTERIOR STRUCTURE

The Earth's Heat Furnace

The Earth's internal heat source provides the energy for our dynamic planet, supplying it with the driving force for plate-tectonic motion, and for on-going catastrophic events such as earthquakes and volcanic eruptions.

This internal heat energy was much greater in the early stages of the Earth than it is today, having accumulated rapidly by heat conversion associated with three separate processes, all of which were most intense during the first few hundred thousand years of the Earth's history: (1) extraterrestrial impacts, (2) gravitational contraction of the Earth's interior, and (3) the radioactive decay of unstable isotopes.

Extraterrestrial Impacts

Most scientists believe that our solar system evolved from the accretion of solid particles derived from a large nebular cloud - the so-called Nebular Hypothesis.

Under this scenario, proto-planet Earth would have grown over time from a barrage of extraterrestrial impacts, increasing its mass with each bombardment. As the proto-planet grew in size its increased gravitational field would have attracted even more objects its surface.

The composition of these colliding bodies would have included metal-rich fragments (i.e.., iron meteorites), rocky fragments (i.e., stony meteorites), and icy fragments (i.e., comets).

Although accretion was much more prevalent in the early stages of the Earth's history, these extraterrestrial collisions are still occurring today, exemplified by shooting stars and fireballs in the night sky, and by the occasional impact of larger bodies on the Earth's surface.

Such particles travel at great velocities, typically ~30,000—50,000 km/hr, similar to that of the Earth as it rotates around the Sun. The very large amount of kinetic energy inherent in these moving bodies is instantly converted to heat energy upon impact, thus providing a component to the Earth's internal heat source.

Gravitational Contraction

In the early stages of planetary accretion, the earth was much less compact than it is today. The accretionary process led to an increasingly greater gravitational attraction, forcing the Earth to contract into a smaller volume. Increased compaction resulted in the conversion of gravitational energy into heat energy, much like a bicycle pump heats up due to the compression of air inside it. Heat conducts very slowly through rock, so that the rapid build up of this heat source within the Earth was not accommodated by an equally rapid loss of heat through the surface.

Decay Of Radioactive Elements

Radioactive elements are inherently unstable, breaking down over time to more stable forms. The unstable isotope Uranium-238, for example, will slowly decay to Lead-206. All such radioactive decay processes release heat as a by product of the on-going reaction. In its early stages of formation, the young earth had a greater complement of radioactive elements, but many of these (e.g., aluminum-26) are short-lived and have decayed to near extinction. Others with a more lengthy rate of decay and are still undergoing this radioactive process, thus still releasing heat energy. The greater complement of unstable elements in the early Earth thus generated a greater amount of heat energy in its initial stages of formation.

Melting and Compositional Differentiation of the Early Earth

The heat buildup inside earth reached a maxim early in the Earth's history and has declined significantly since. The greater heat content of the early Earth was the product of (1) a greater abundance of radioactive elements, (2) a greater number of impacts, and (3) the early gravitational crowding. The initial accretion of particles resulted in a rather homogeneous sphere composed of a loose amalgam of metallic fragments (iron meteorites), rocky fragments (stony meteorites), and icy fragments (comets). However, the increased heat content of the early Earth resulted in melting of the Earth's interior, so that the young planet became density stratified with the heavier (metallic) materials sinking to the center of the earth, and the lighter (rocky) materials floating upward toward the surface of the earth. The very lightest volatile materials (derived from comets) were easily melted or vaporized, rising beyond the earth's rocky surface to form the early oceans and the atmosphere. We now have a differentiated earth due to melting and mobilization of materials driven by the earth's internal heat engine. This has resulted in the development of a series of concentric layers that are both density and compositionally stratified. This demonstrated in the diagram below, *courtesy of the USGS.*

Within the mantle exists the asthenosphere (Grk. asthenos = weak), between about 100 km and 350 km, which is a special zone composed of hot, weak material that is capable of gradual flow.

The layer above the asthenosphere is the lithosphere (Grk. lithos = rock), the rigid and relatively cool outer layer of the earth, composed of both crust and a portion of the upper mantle.

Lying above the lithosphere is (1) the liquid hydrosphere, comprising 71% of the Earth's surface, and (2) that the still lighter gaseous atmosphere, both of which were ultimately derived from the accretion of comets. The occurrence of these volatile components along the outermost portion of the Earth is a product of volcanic outgassing during the differentiation event.

DYNAMICS OF GEOMORPHOLOGY

To place geomorphology upon sound foundations for quantitative research into fundamental principles, it is proposed that geomorphic processes be treated as gravitational or molecular shear stresses acting upon elastic, plastic, or fluid earth materials to produce the characteristic varieties of strain, or failure, that constitute weathering, erosion, transportation and deposition.

Shear stresses affecting earth materials are here divided into two major categories: gravitational and molecular. Gravitational stresses activate all downslope movements of matter, hence include all mass movements, all fluvial and glacial processes. Indirect gravitational stresses activate wave-and tide-induced currents and winds. Phenomena of gravitational shear stresses are subdivided according to behavior of rock, soil, ice, water, and air as elastic or plastic solids and viscous fluids. The order of classification is generally that of decreasing internal resistance to shear and, secondarily, of laminar to turbulent flow.

Molecular stresses are those induced by temperature changes, crystallization and melting, absorption and desiccation, or osmosis. These stresses act in random or unrelated directions with respect to gravity. Surficial creep results from combination of gravitational and molecular stresses on a slope. Chemical processes of solution and acid reaction are considered separately.

A fully dynamic approach requires analysis of geomorphic processes in terms of clearly defined open systems which tend to achieve steady states of operation and are self-regulatory to a large degree. Formulation of mathematical models, both by rational deduction and empirical analysis of observational data, to relate energy, mass, and time is the ultimate goal of the dynamic approach.

Geomorphology

Geomorphology is the scientific study of landforms and the processes that shape them. Geomorphologists seek to understand why landscapes look the way they do, to understand landform history and dynamics, and to predict future changes through a combination of field observations, physical experiments, and numerical modeling. Geomorphology is practiced within geography, geology, geodesy, engineering geology, archaeology, and geotechnical engineering, and this broad base of interest contributes to a wide variety of research styles and interests within the field.

Overview

The surface of Earth is modified by a combination of surface processes that sculpt landscapes and geologic processes that cause tectonic uplift and subsidence. Surface processes comprise the action of water, wind, ice, fire, and living things on the surface of the Earth, along with chemical reactions that form soils and alter material properties, the stability and rate of change of topography under the force of gravity, and other factors, such as (in the very recent past) human alteration of the landscape. Many of these factors are strongly mediated by climate. Geologic forcings include the uplift of mountain ranges, the growth of volcanoes, isostatic changes in land surface elevation (sometimes in response to surface processes), and the formation of deep sedimentary basins where the surface of Earth drops and is filled with material eroded from other parts of the landscape. The Earth surface and its topography therefore are an intersection of climatic, hydrologic, and biologic action with geologic processes.

The broad-scale topographies of Earth illustrate this intersection of surface and subsurface action. Mountain belts are uplifted due to geologic processes. Denudation of these high uplifted regions produces sediment that is transported and deposited elsewhere within the landscape or off the coast. On progressively smaller scales, similar ideas apply, where individual landforms evolve in response to the balance of additive processes (uplift and deposition) and subtractive processes (subsidence and erosion).

Often, these processes directly affect each other: ice sheets, water, and sediment are all loads that change topography through flexural isostasy. Topography can modify the local climate, for example through orographic precipitation, which in turn modifies the topography by changing the hydrologic regime in which it evolves. Many geomorphologists are particularly interested in the potential for feedbacks between climate and tectonics mediated by geomorphic processes.

In addition to these broad-scale questions, geomorphologists address issues that are more specific and/or more local. Glacial geomorphologists investigate glacial deposits such as moraines, eskers, and proglacial lakes, as well as glacial erosional features, to build chronologies of both small glaciers and large ice sheets and understand their motions and effects upon the landscape. Fluvial geomorphologists focus on rivers, how they transport sediment, migrate across the landscape, cut into bedrock, respond to environmental and tectonic changes, and interact with humans.

Soils geomorphologists investigate soil profiles and chemistry to learn about the history of a particular landscape and understand how climate, biota, and rock interact. Other geomorphologists study how hillslopes form and change. Still others investigate the relationships between ecology and geomorphology. Because geomorphology is defined to comprise everything related to the surface of Earth and its modification, it is a broad field with many facets.

Practical applications of geomorphology include hazard assessment (such as landslide prediction and mitigation), river control and stream restoration, and coastal protection.

History

With some notable exceptions, geomorphology is a relatively young science, growing along with interest in other aspects of the Earth Sciences in the mid 19th century. This section provides a very brief outline of some of the major figures and events in its development.

Ancient Geomorphology

Perhaps the earliest one to devise a theory of geomorphology was the polymath Chinese scientist and statesman Shen Kuo (1031-1095 AD). This was based on his observation of marine fossil shells in a geological stratum of a mountain hundreds of miles from the Pacific Ocean. Noticing bivalve shells running in a horizontal span along the cut section of a cliffside, he theorized that the cliff was once the pre-historic location of a seashore that had shifted hundreds of miles over the centuries. He inferred that the land was reshaped and formed by soil erosion of the mountains and by deposition of silt, after observing strange natural erosions of the Taihang Mountains and the Yandang Mountain near Wenzhou. Furthermore, he promoted the theory of gradual climate change over centuries of time once ancient petrified bamboos were found to be preserved underground in the dry, northern climate zone of *Yanzhou*, which is now modern day Yan'an, Shaanxi province.

Early Modern Geomorphology

The first use of the word geomorphology was likely to be in the German language when it appeared in Laumann's 1858 work. Keith Tinkler has suggested that the word came into general use in English, German and French after John Wesley Powell and W. J. McGee used it in the International Geological Conference of 1891.

An early popular geomorphic model was the *geographical cycle* or the *cycle of erosion*, developed by William Morris Davis between 1884 and 1899. The cycle was inspired by theories of uniformitarianism first formulated by James Hutton (1726–1797). Concerning valley forms, uniformitarianism depicted the cycle as a sequence in which a river cuts a valley more and more deeply, but then erosion of side valleys eventually flatten the terrain again, to a lower elevation.

Tectonic uplift could start the cycle over. Many studies in geomorphology in the decades following Davis' development of his theories sought to fit their ideas into this framework for broad scale landscape evolution, and are often today termed "Davisian". Davis' ideas have largely been superseded today, mainly due to their lack of predictive power and qualitative nature, but he remains an extremely important figure in the history of the subject.

In the 1920s, Walther Penck developed an alternative model to Davis', believing that landform evolution was better described as a balance between ongoing processes of uplift and denudation, rather than Davis' single uplift followed by decay. However, due to his relatively young death, disputes with Davis and a lack of English translation of his work his ideas were not widely recognised for many years.

These authors were both attempting to place the study of the evolution of the Earth's surface on a more generalized, globally relevant footing than had existed before. In the earlier parts of the 19th century, authors-especially in Europe-had tended to attribute the form of landscape to local climate, and in particular to the specific effects of glaciation and periglacial processes. In contrast, both Davis and Penck were seeking to emphasize the importance of evolution of landscapes through time and the generality of Earth surface processes across different landscapes under different conditions.

Quantitative Geomorphology

While Penck and Davis and their followers were writing and studying primarily in Western Europe, another, largely separate, school of geomorphology was developed in the United States in the middle years of the 20th century.

Following the early trailblazing work of Grove Karl Gilbert around the turn of the 20th century, a group of natural scientists, geologists and hydraulic engineers including Ralph Alger Bagnold, John Hack, Luna Leopold, Thomas Maddock and Arthur Strahler began to research the form of landscape elements such as rivers and hillslopes by taking systematic, direct, quantitative measurements of aspects of them and investigating the scaling of these measurements. These methods began to allow prediction of the past and future behavior of landscapes from present observations, and were later to develop into what the modern trend of a highly quantitative approach to geomorphic problems. Quantitative geomorphology can involve fluid dynamics and solid mechanics, geomorphometry, laboratory studies, field measurements, theoretical work, and full landscape evolution modeling. These approaches are used to understand weathering and the formation of soils, sediment transport, landscape change, and the interactions between climate, tectonics, erosion, and deposition.

SOCIAL CONSTRUCTIVISM OR RELATIVISM

Perhaps the strongest blow to logical positivism was the recognition that the methods of science, including observation, are inherently theory laden (Hanson 1958; Kuhn 1970). "Observation is always observation in the light of theories". Because an observation or observation statement can be wrong, it is hypothetical. A theory-neutral form of observation or a significant ontological distinction between the observable and the unobservable is untenable (Maxwell 1962; Putnam 1962). Extreme versions of the theory—ladenness premise provide the basis for relativistic perspectives on scientific knowledge—that is, the epistemological view that the acceptability or unacceptability of scientific claims to knowledge is not objective but depends on social consensus in a specific group of scientists. If observations are strongly dependent on the theories they are used to test, theories are incommensurable because scientists holding different theories will observe different phenomena (Hanson 1958, 4-10), which thereby undermines the adjudicatory power normally accorded to observation.

Moreover, relativists argue that the incommensurability thesis applies not only to modern science but also to the history of science. Based on historical analysis, T. S. Kuhn (1970) and P. K. Feyerabend (1975) concluded that radical conceptual shifts sometimes occur wherein the meaning of scientific terms is not preserved between successive theories or paradigms.

Thus scientists operating under an old theory cannot communicate with those adhering to a new one. During scientific revolutions a complete reconceptualisation of reality occurs that is regulated primarily by social consensus in the scientific community. Science does not progress towards truth; it merely changes its communal perspective on reality. This viewpoint greatly threatens the notion that science reveals truths or theory-independent facts about an objective reality. The aim of science is to solve problems, not to seek the truth.

This relativist perspective did not gain a strong foothold in the United States, but it flourished abroad, specifically at the University of Edinburgh, where it became the foundation of social constructivism (Giere 1988, 50-61). Strong social constructivists (Bloor 1976; Barnes 1982) maintain that theory acceptance and theory change can be explained entirely by social factors. By contrast, weak constructivists (Kuhn 1977; Longino 1990) recognise that other factors have a role in these processes but assert that social factors dominate.

The social-constructivist approach to the study of science is heavily naturalistic, emphasising empirical analysis of actual scientific practice to develop causal explanations of how scientists produce knowledge.

Most of these participant-observer investigations emphasise how social negotiations among scientists play a major role in new claims to knowledge and in the fixation of belief. New scientific knowledge is thus socially constructed in the sense that it is mainly a product of social factors rather than of interaction with the natural world (Collins 1981). Opponents contend that if this assertion be true, the approach is self-refuting because its advocates are using the methods of science to justify the validity of the approach (Niiniluoto 1991a).

POSTPOSITIVIST EMPIRICISM

Although contemporary forms of metaphysical instrumentalism exist (Stegmuller 1976), other postpositivist forms of empiricism reject a strict instrumentalist view of scientific theories (van Fraassen 1980). According to the later perspective, theoretical statements do not acquire meaning only through their connection with the observable, but instead are to be construed literally. They have meaning in and of themselves and are capable of being true or false.

By accepting that statements about unobservable phenomena have truth value, this type of empiricism avoids the dilemma posed by the lack of a clear ontological distinction between the observable and the unobservable. However, it does maintain that directly observable phenomena have privileged epistemological status. In other words, knowledge about directly observable phenomena has a more certain evidential basis than do claims to knowledge about unobservable phenomena. This type of empiricism has been characterised as epistemological instrumentalism.

Although epistemological instrumentalism acknowledges that the theoretical entities in theories are best thought of as potentially real, it is still antirealist in that it maintains that "good" scientific theories do not have to be true; thus acceptance of a theory does not mandate belief in its truth or in the reality of the unobservable entities it postulates. In other words, the truth value of a theory is not the basis for theory choice. Instead, the aim of a science is to generate theories that are empirically adequate (van Fraassen 1980) or that solve problems (Laudan 1977, 1984). Acceptance of a theory commits scientists only to the belief that the theory is empirically adequate or solves problems effectively, not to the position that it is true. Empirical adequacy refers to a high degree of conformity between directly observable phenomena and what a theory predicts about these observable aspects of the world.

Epistemological instrumentalism is strongly empiricist in that it involves a search for truth only about observable phenomena and maintains that science provides no means to justify belief in theoretical entities. According to the contemporary empiricists, ultimately the only basis for evaluating theoretical hypotheses is observational evidence. Because all theoretical statements extend beyond this observational evidence, they are underdetermined in the sense that they are less probable than the empirical evidence. Observational evidence cannot determine the truth of theoretical propositions, and belief in the truth of such statements always involves greater epistemic risk than does belief in the truth of statements about the observable (van Fraassen 1985). Although this form of empiricism assigns privileged epistemological status to statements about observable phenomena, it fails to provide a definitive philosophical argument to support this status.

SCIENTIFIC REALISM

Contemporary scientific realism derives from philosophical ideas espoused by Charles S. Peirce near the end of the nineteenth century and by Karl Popper since 1930 (Niiniluoto 1991b). Peirce and Popper both interpreted scientific knowledge as corrigible and rejected the received-view notion of naive realism that science progresses through the steady accumulation of verified truths about the observable world. Nevertheless, both argued that truth plays an important role in scientific progress because over time, according to Peirce, science moves towards the truth about reality or, according to Popper, develops theories that are more and more truthlike. Popper (1968, 423-425) also denied metaphysical instrumentalism and embraced a realist interpretation of scientific theories.

Numerous versions of realism exist in contemporary philosophy of science. These variations can be divided into three categories: internal realism, entity realism, and critical realism (Niiniluoto 1991b). Internal realists reject the view of truth as correspondence between statements about the world and facts about an objective, accessible reality. Instead, they replace this definition with a pragmatic, epistemic notion of truth, such as ideal acceptability or eventual perfection of scientific explanations (Newton-Smith 1989).

Entity realists subscribe to the correspondence theory of truth, but they accept existence claims only about unobservable entities that can be manipulated experimentally to produce detectable effects, such as electrons, neutrinos, or positrons, and deny a realist interpretation of theoretical laws.

Although these realist philosophical positions have merit, the label scientific realism is normally associated with critical realism, which most commonly serves as the focal point for opposing philosophical criticism by social constructivists and postpositivist empiricists. Critical realists maintain four assumptions. Terms in scientific theories that describe unobservable entities or causal mechanisms should be interpreted realistically: the terms refer to phenomena that genuinely exist. Reality exists independently of human thought, and the description of this reality embodied in scientific theories is largely independent of theoretical commitments. The methods of science provide a means for determining the approximate truth of all aspects of scientific theories, including statements about unobservables.

The historical progress of science consists of the development of successive theories that become more truthlike over time, with later theories building on the knowledge embodied in previous ones; thus current scientific theories in mature sciences are at least approximately true (Boyd 1983; Niiniluoto 1984). This perspective is a strong version of realism in that it includes ontological and epistemological components. Clearly, critical realism adopts the perspective that the aim of science is to seek the truth, not merely to solve problems. Critical realists readily acknowledge that all aspects of scientific enquiry are theory laden. They maintain that this characteristic of science permits knowledge of a theory-independent world and accounts for the reliability of this knowledge. This assertion is supported by an abductive argument, or inference to the best explanation, which has the following structure.

- Premise 1—If scientific theories are approximately true, they will be empirically successful.
- Premise 2—Many scientific theories are empirically successful.
- Probable conclusion—Many scientific theories are approximately true.

In other words, realists argue that the only plausible explanation for the overwhelming empirical success of a heavily theory-dependent enterprise like science is that the theories on which it is based are at least approximately true. This line of reasoning is known as the no-miracles argument: if the best current theories are not at least approximately true, the empirical success of modern science must be viewed as a miracle.

According to critical realists, current theories are approximately true and are the foundation for scientific progress. Although an infinite number of possible explanations exists for any anomalous empirical result, scientists do not assign equal weight to each explanation. Instead, based on their current theoretical knowledge, they immediately discount all but a few: existing theory plays an evidential role in choosing between competing theories and serves as a basis for the development of new theories (Boyd 1983). Theoretical knowledge is evidentially as important as, if not more important than, empirical knowledge. Also, the theory ladenness of observation is not problematic because it is the true or approximately true theoretical knowledge on which scientific observational procedures are based that allows access to new theoretical knowledge about theory-independent reality.

This situation is best exemplified when scientists use theory to develop sophisticated measurement techniques or instruments that permit detection of the effects of unobservable entities and causal mechanisms (Brown 1990). The epistemological component of critical realism requires justification in light of the constructivist argument that science occasionally undergoes radical conceptual shifts in which previous theories are completely abandoned, with resulting semantic incommensurability between scientific theories.

The notion of radical conceptual shifts greatly threatens the realist image of scientific progress. For current theories to be approximately true, they must preserve as limiting cases antecedent theories that were also approximately true. However, if successive theories are truly incommensurable, this conception of scientific progress is difficult to defend. To address this problem, realists adopt a causal or referential theory of meaning wherein the meaning of a theoretical term that describes an unobservable entity is determined not by the properties of the entity, which depend on the theory, but by an introducing event that puts scientists in the appropriate causal relationship with the entity (Putnam 1973). Usually this appropriate relationship involves a situation wherein the entity is present and produces certain effects. The properties assigned to the entity may differ between theories or over time as theories change, but as long as scientists agree that the entity responsible for certain effects under theory A is the same entity responsible for effects under theory B, reference and meaning are preserved. The development of true definitions of scientific phenomena based on assignation of properties is an a posteriori theoretical issue rather than an a priori problem of convention (Boyd 1990).

Critical realism also rejects metaphysical realism, the notion that reality consists of facts that exist independently of human conceptualisation and that one true theory is capable of describing this reality. Instead, critical realists acknowledge that facts are always created by humans in the sense that they are in part linguistic entities that presuppose the existence of human minds and conceptual frameworks (Niiniluoto 1984, 177-178). However, in contrast to constructivists, critical realists maintain that even though language is required to carve up the world into facts that become the object of scientific investigations, the truth of statements about these facts is determined not by the conceptual framework but by the causal interaction with an objective reality. This view of science is fundamentally nonreductionist because it implies that no single language or science has privileged status concerning access to reality and that unrelated conceptual frameworks can reveal different truths about reality (Boyd 1989).

PHYSICAL CONDITIONS OF THE EARTH'S INTERIOR

Three centuries ago, the English scientist Isaac Newton calculated, from his studies of planets and the force of gravity, that the average density of the Earth is twice that of surface rocks and therefore that the Earth's interior must be composed of much denser material. Our knowledge of what's inside the Earth has improved immensely since Newton's time, but his estimate of the density remains essentially unchanged. Our current information comes from studies of the paths and characteristics of earthquake waves travelling through the Earth, as well as from laboratory experiments on surface minerals and rocks at high pressure and temperature. Other important data on the Earth's interior come from geological observation of surface rocks and studies of the Earth's motions in the Solar System, its gravity and magnetic fields, and the flow of heat from inside the Earth.

The planet Earth is made up of three main shells: the very thin, brittle crust, the mantle, and the core; the mantle and core are each divided into two parts. All parts are drawn to scale on the cover of this publication, and a table at the end lists the thicknesses of the parts. Although the core and mantle are about equal in thickness, the core actually forms only 15 per cent of the Earth's volume, whereas the mantle occupies 84 per cent. The crust makes up the remaining 1 per cent. Our knowledge of the layering and chemical composition of the Earth is steadily being improved by earth scientists doing laboratory experiments on rocks at high pressure and analyzing earthquake records on computers.

The Crust

Because the crust is accessible to us, its geology has been extensively studied, and therefore much more information is known about its structure and composition than about the structure and composition of the mantle and core. Within the crust, intricate patterns are created when rocks are redistributed and deposited in layers through the geologic processes of eruption and intrusion of lava, erosion, and consolidation of rock particles, and solidification and recrystallization of porous rock.

By the large-scale process of plate tectonics, about twelve plates, which contain combinations of continents and ocean basins, have moved around on the Earth's surface through much of geologic time. The edges of the plates are marked by concentrations of earthquakes and volcanoes. Collisions of plates can produce mountains like the Himalayas, the tallest range in the world.

The plates include the crust and part of the upper mantle, and they move over a hot, yielding upper mantle zone at very slow rates of a few centimeters per year, slower than the rate at which fingernails grow. The crust is much thinner under the oceans than under continents. The boundary between the crust and mantle is called the Mohorovicic discontinuity; it is named in humour of the man who discovered it, the Croatian scientist

Andrija Mohorovicic. No one has ever seen this boundary, but it can be detected by a sharp increase downward in the speed of earthquake waves there. The explanation for the increase at the Moho is presumed to be a change in rock types. Drill holes to penetrate the Moho have been proposed, and a Soviet hole on the Kola Peninsula has been drilled to a depth of 12 kilometers, but drilling expense increases enormously with depth, and Moho penetration is not likely very soon.

The Mantle

Our knowledge of the upper mantle, including the tectonic plates, is derived from analyses of earthquake waves; heat flow, magnetic, and gravity studies; and laboratory experiments on rocks and minerals. Between 100 and 200 kilometers below the Earth's surface, the temperature of the rock is near the melting point; molten rock erupted by some volcanoes originates in this region of the mantle. This zone of extremely yielding rock has a slightly lower velocity of earthquake waves and is presumed to be the layer on which the tectonic plates ride.

This low-velocity zone is a transition zone in the upper mantle; it contains two discontinuities caused by changes from less dense to more dense minerals. The chemical composition and crystal forms of these minerals have been identified by laboratory experiments at high pressure and temperature. The lower mantle, below the transition zone, is made up of relatively simple iron and magnesium silicate minerals, which change gradually with depth to very dense forms. Going from mantle to core, there is a marked decrease in earthquake wave velocity and a marked increase in density.

The Core

The paths curve because the different rock types found at different depths change the speed at which the waves travel. Solid lines marked P are compressional waves; dashed lines marked S are shear waves. S waves do not travel through the core but may be converted to compressional waves on entering the core. Waves may be reflected at the surface.

The core was the first internal structural element to be identified. It was discovered in 1906 by R.D. Oldham, from his study of earthquake records, and it helped to explain Newton's calculation of the Earth's density. The outer core is presumed to be liquid because it does not transmit shear (S) waves and because the velocity of compressional (P) waves that pass through it is sharply reduced. The inner core is considered to be solid because of the behaviour of P and S waves passing through it.

The Structure of the Moon

The Moon, our fellow-traveller in space, has a diameter half that of the Earth's core, and it revolves around the Earth, as all the planets revolve around the Sun, under the force of gravity.

Moonquakes of very low energy are caused by land tides produced by the pull of Earth's gravity, and, from analysis of moonquake data, scientists believe the Moon has two layers: a crust, from the surface to 65 kilometers depth, and an inner, more dense mantle from the crust to the center at 3,700 kilometers.

The crust is presumed to be com-posed primarily of rocks containing feldspar, calcium aluminum silicate, and lesser pyrox-ene, iron and magnesium silicate; the crust also contains basalt in the mares, which contains less iron and more titanium than earth basalt. The mantle is thought to be made up of calcic peridotite, containing both pyroxene and feldspar.

6

Oceanography

Oceanography has been aptly defined as the study of the world below the surface of the sea: it should include the contact zone between sea and atmosphere. Present–day acceptance it has to do with all the characteristics of the bottom and margins of the sea, of the sea water, and of the inhabitants of the latter. Thus widely combining geophysics, geochemistry, and biology, it is inclusive, as is, of course, characteristic of any 'young' science: and modern oceanography is in its youth. But in this case it is not so much immaturity that is responsible for the fact that these several subsciences are still grouped together, but rather the realization that the physics, chemistry, and biology of the sea water are not only important per se, but that in most of the basic problems of the sea all three of these subdivisions have a part. And with every advance in our knowledge of the sea making this interdependence more and more apparent, it is not likely, that we shall soon see any general abandon ment of this concept of oceanography as a mother science, the branches of which, though necessarily attacked by different disciplines, are intertwined too closely to be torn apart. Every oceanic biologist should, therefore, be grounded in the principles of geophysics and geochemistry; every chemical or physical oceanographer in some of the oceanic aspects of biology.

A feature equally inherent in sea science is that it is no less inclusive from the geographic standpoint, because the subjects with which it is concerned cover so large a part of the surface of our planet.. And the vastness of the areas to be considered, whatever phase of the sea be under consideration, has; determined the paths that the science of oceanography has followed in its advance from, its early beginnings to its present state. Most oceanographers, too, would agree that the geographic factor has likewise been responsible for a failure to progress at a rate commensurate either with the relative importance, of this field of knowledge in the general household of science, or with the amount of energy that has been, devoted thereto during the past quarter-century. In the nature of things, the oceanographer constantly meets a, twofold obstacle of another sort, when he attempts to extend his investigations out from the shore–line to the high seas, no matter from what headquarters he may work, in the necessity of studying the majority of oceanic phenomena and events within the sea, not merely upon it, or from its borders. Even if his investigation be of a sort that can be carried on in a laboratory on shore, the raw data must be gathered at sea. Therefore he must have a boat, a necessity that places him at a disadvantage as compared with the general biologist who turns to marine animals chiefly for convenience, and so can pick up many things of interest on a stroll along tideline.

If the student is to venture out more than a few miles from land, his craft must be large enough to contain living quarters and to navigate safely in all weathers, for oceanography is impossible unless some one does go out to sea, on short trips or on long. That is to say, for even one investigator, or one party, to gather information ofany kind about the ocean in appreciable amount demands the labours of many, as reflected in the maintenance and operation of a seagoing craft, with crew to man her, with supplies for their subsistence; also with fuel for her propulsion. And as any craft larger than a rowboat is an expensive means of conveyance for a small number of passengers, it follows that exploration into the economy of the high seas is essentially a costly undertaking.

The expense of extended voyages, combined with the necessity for such if large sea areas were to be studied more intensively than could be done from examination of vessel's log books, was no doubt the chief reason that systematic examination, even of the surface of the sea, was not seriously undertaken until the middle of the nineteenth century. But when, at about this period, science awoke to the whole new world for exploration that was offered by the oceans, it was soon learned that no very serious technical difficulties, apart from expense, were involved in extending investigations down into the abyss, whether it was a question of developing the contour of the sea. floor, of gathering samples of the bottom, of sampling the living creatures, or of measuring the physical and chemical characteristics of the water.

It would, indeed, have been quite within the technical abilities of the Romans of Pliny's day to have plumbed the depths of the Mediterranean and to have explored its deep-water biota, though of course examination of the temperature and salinity of the sea must, in any case, have awaited the development of the sciences of physics and chemistry as we now know them. Efficient gear was, in fact, so rapidly developed in the three years

1868–70, as soon as a serious start in that direction was made on the voyages of the 'Lightning' and of the 'Porcupine,' that when the 'Challenger' sailed two years later on the first oceanwide exploration. of the deeps, the scientists on board were already in position efficiently to reap the rich harvest that stimulated many other subsequent expeditions in various parts of the world. From that time forward, with every fresh venture below the surface of the sea, and with constant improvement in technical methods of operating gear of different kinds at great depths, a flood of new facts came pouring in so rapidly that more was learned about the,sea and about the inhabitants of its deeps during the last thirty years of the nineteenth century than had been up till then. This was the heyday of the deep-sea exploring expedition, when one cruise after another was sent out by different maritime nations, when the broad relief of the ocean floors was mapped, when the general nature of the submarine sediments was determined, when the distribution of temperature and salinity was worked 'out in its essential outlines over the high seas, and when the general characteristics of the deep–sea fauna were explored.

While this regional-descriptive era of oceanography will never definitely close so long as the science of the sea is pursued, there came a change, towards the end of the century just past, when persistence in the old discursive methods, determined by established habits of thought, no longer yielded new and wonderful discoveries at the rate that had been the order of the day when no one knew what was to be found at the bottom of the sea.

Thenceforth, with increasing frequency, continued exploration along these preliminary lines yielded results more corroborative than novel. And a period of general oceanographic stagnation might then have succeeded to the preceding peak of activity had there not arisen new schools, centering their attention on the biologic economy of the inhabitants of the ocean as related to their physical–chemical environment, on mathematical analysis of the internal dynamics of the sea water, and on the geologic bearing of submarine topography and sedimentation, rather than on areal surveys of one or another feature of the sea. This conscious alteration of viewpoint, from the descriptive to the analytic, is one of two chief factors that gives to oceanography its present tone: the other is the growth of an economic demand that oceanography afford practical assistance to the sea fisheries.

This demand developed first in northwestern Europe, where, as it chanced, the fisheries were so rapidly expanding, and increasing in intensity through the adoption of more effective methods of fishing, that dread of depletion began to loom in the offing, just when oceanography was approaching the end of its nineteenth–century boom; *i.e.*, just when it needed a fresh stimulus.

The immediate, practical result was a concentration of attention on limited coastwise areas as contrasted with the broad oceans, and the development of an international and official organization—he Conseil International pour l'Exploration de la Mer–with power to coordinate the scientific efforts of the Fisheries Bureaus of the several nations fronting on these areas in northwestern Europe.

It is an interesting speculation whether, without this enforced direction of scientific attention to the North Sea region, we should have come to appreciate, as clearly as we do today, that application to the adjacent oceans of principles established by intensive investigation of such test cases offers the most promising lines of approach to many of the broad, underlying problems of oceanography.

However, this may be, we can hardly doubt that the advance of oceanography on the analytic-synthetic side would have continued slow and halting had not the Conseil and other coordinating institutions of more recent birth, but with similar aims, added their unifying influence to the attempts at synthetic investigation that would in any case have followed the alteration in viewpoint. And it is certain that, today, the most rapid approaches towards an understanding of events in the sea are being made by orderly, intensive, and concerted attacks upon one or another phase, via definitely stated and apparently illustrative problems, rather than by haphazard accumulation of unrelated facts, gathered in the hope that somehow, sometime, these may be fitted together by some one.

This is reflected in the fact that several broad–scale expeditions that have been sent out within the last few years—'Meteor,' 'Dana,' 'Carnegie,' 'Marion'—have devoted their attention chiefly to extending to the high seas special lines of investigation the theoretical basis for which had already been developed from intensive studies falling in the general category just stated.

The foregoing remarks are introductory to the thesis that a discussion of certain of the underlying problems that seem most clearly to illustrate the general fields of research falling within the province of the oceanographer, and that are now most to the fore, is integral in any rational exposition of the scope and present status of this inclusive branch of science.

To list all the problems that await the oceanographer will never be possible so long as science lives, for new ones will constantly unfold, as the boundaries of knowledge are rolled back.

In practice oceanography falls most conveniently into three chief divisions:

1. The geological;
2. The physical-chemical;
3. The biological.

To consider first the problems of the shape and composition of the basins that hold the oceans; next, those associated with the physical character and chemical composition of the waters that fill these basins; and third, those of the nature and activities of the animals and plants that inhabit the waters is therefore a rational order of presentation. Subsequent stages discuss the fundamental unity of these different divisions, and outline certain of the direct economic benefits that may be expected to accrue from the study of oceanography.

A word is perhaps due the reader to explain our omission of any references to authorities. Citations are an essential part of any presentation of the results of investigation, or of any textbook, but this is not necessarily true when the discussion is of things to be studied. In the present case to have mentioned authors would have necessitated a general bibliography of the subject; also roster of oceanographers, for each of the latter has by his published work helped to make this book possible. To cut this Gordian knot personal references are omitted.

History

Humans first acquired knowledge of the waves and currents of the seas and oceans in pre–historic times. Observations on tides are recorded by Aristotle and Strabo. Early modern exploration of the oceans was primarily for cartography and mainly limited to its surfaces and of the creatures that fishermen brought up in nets, though depth soundings by lead line were taken. Although Juan Ponce de Leon in 1513 first identified the Gulf Stream, and the current was well-known to mariners, Benjamin Franklin made the first scientific study of it and gave it its name. Franklin measured water temperatures during several Atlantic crossings and correctly explained the Gulf Stream's cause. Franklin and Timothy Folger printed the first map of the Gulf Stream in 1769–1770.

When Louis Antoine de Bougainville, who voyaged between 1766 and 1769, and James Cook, who voyaged from 1768 to 1779, carried out their explorations in the South Pacific, information on the oceans themselves formed part of the reports. James Rennell wrote the first scientific textbooks about currents in the Atlantic and Indian oceans during the late 18th and at the beginning of 19th century. Sir James Clark Ross took the first modern sounding in deep sea in 1840, and Charles Darwin published a document on reefs and the formation of atolls as a result of the second voyage of HMS Beagle in 1831-6. Robert FitzRoy published a report in four volumes of the three voyages of the Beagle. In 1841–1842 Edward Forbes undertook dredging in the Aegean Sea that founded marine ecology.

As first superintendent of the United States Naval Observatory (1842–1861) Matthew Fontaine Maury devoted his time to the study of marine meteorology, navigation, and charting prevailing winds and currents. His Physical Geography of the Sea, 1855 was the first textbook of oceanography. Many nations sent oceanographic observations to Maury at the Naval Observatory, where he and his colleagues evaluated the information and gave the results worldwide distribution.

The steep slope beyond the continental shelves was discovered in 1849. The first successful laying of transatlantic telegraph cable in August 1858 confirmed the presence of an underwater "telegraphic plateau" mid–ocean ridge. After the middle of the 19th century, scientific societies were processing a flood of new terrestrial botanical and zoological information.

In 1871, under the recommendations of the Royal Society of London, the British government sponsored an expedition to explore world's oceans and conduct scientific investigations. Under that sponsorship the Scots Charles Wyville Thompson and Sir John Murray launched the Challenger expedition (1872–1876). The results of this were published in 50 volumes covering biological, physical and geological aspects. 4417 new species were discovered. Other European and American nations also sent out scientific expeditions (as did private individuals and institutions). The first purpose built oceanographic ship, the "Albatros" was built in 1882. The four-month 1910 North Atlantic expedition headed by Sir John Murray and Johan Hjort was at that time the most ambitious research oceanographic and marine zoological project ever, and led to the classic 1912 book The Depths of the Ocean.

Oceanographic institutes dedicated to the study of oceanography were founded. In the United States, these included the Scripps Institution of Oceanography in 1892, Woods Hole Oceanographic Institution in 1930, Virginia Institute of Marine Science in 1938, Lamont-Doherty Earth Observatory at Columbia University, and the School of Oceanography at University of Washington. In Britain, there is a major research institution: National Oceanography Centre, Southampton which is the successor to the Institute of Oceanography. In Australia, CSIRO Marine and Atmospheric Research, known as CMAR, is a leading center. In 1921 the International Hydrographic Bureau (IHB) was formed in Monaco.

In 1893, Fridtjof Nansen allowed his ship "Fram" to be frozen in the Arctic ice. As a result he was able to obtain oceanographic data as well as meteorological and astronomical data. The first international organization of oceanography was created in 1902 as the International Council for the Exploration of the Sea.

The first acoustic measurement of sea depth was made in 1914. Between 1925 and 1927 the "Meteor" expedition gathered 70,000 ocean depth measurements using an echo sounder, surveying the Mid atlantic ridge. The Great Global Rift, running along the Mid Atlantic Ridge, was discovered by Maurice Ewing and Bruce Heezen in 1953 while the mountain range under the Arctic was found in 1954 by the Arctic Institute of the USSR. The theory of seafloor spreading was developed in 1960 by Harry Hammond Hess. The Ocean Drilling Project started in 1966. Deep sea vents were discovered in 1977 by John Corlis and Robert Ballard in the submersible "Alvin".

In the 1950s, Auguste Piccard invented the bathyscaphe and used the "Trieste" to investigate the ocean's depths. The nuclear submarine Nautilus made the first journey under the ice to the North Pole in 1958. In 1962 there was the first deployment of FLIP (Floating Instrument Platform), a 355 foot spar buoy. Then, in 1966, the U.S. Congress created a National Council for Marine Resources and Engineering Development. NOAA was put in charge of exploring and studying all aspects of Oceanography in the USA. It also enabled the National Science Foundation to award Sea Grant College funding to multi–disciplinary researchers in the field of oceanography.

From the 1970s, there has been much emphasis on the application of large scale computers to oceanography to allow numerical predictions of ocean conditions and as a part of overall environmental change prediction. An oceanographic buoy array was established in the Pacific to allow prediction of El Nino events. 1990 saw the start of the World Ocean Circulation Experiment (WOCE) which continued until 2002. Geosat seafloor mapping data became available in 1995. In 1942, Sverdrup and Fleming published "The Ocean" which was a major landmark. "The Sea" edited by M.N. Hill was published in 1962 while the "Encyclopedia of Oceanography" by Rhodes Fairbridge was published in 1966.

Connection to the atmosphere

The study of the oceans is linked to understanding global climate changes, potential global warming and related biosphere concerns. The atmosphere and ocean are linked because of evaporation and precipitation as well as thermal flux (and solar insolation). Wind stress is a major driver of ocean currents while the ocean is a sink for atmospheric carbon dioxide.

- Our planet is invested with two great oceans; one visible, the other invisible; one underfoot, the other overhead; one entirely envelopes it, the other covers about two thirds of its surface.

Branches

The study of oceanography is divided into branches:

- Biological oceanography, or marine biology, is the study of the plants, animals and microbes of the oceans and their ecological interaction with the ocean;
- Chemical oceanography, or marine chemistry, is the study of the chemistry of the ocean and its chemical interaction with the atmosphere;
- Geological oceanography, or marine geology, is the study of the geology of the ocean floor including plate tectonics;
- Physical oceanography, or marine physics, studies the ocean's physical attributes including temperature-salinity structure, mixing, waves, internal waves, surface tides, internal tides, and currents. Of particular interest is the behaviour of sound (acoustical oceanography), light (optical oceanography) and radio waves in the ocean.

These branches reflect the fact that many oceanographers are first trained in the exact sciences or mathematics and then focus on applying their interdisciplinary knowledge, skills and abilities to oceanography. Data derived from the work of Oceanographers is used in marine engineering, in the design and building of oil platforms, ships, harbours, and other structures that allow us to use the ocean safely. Oceanographic data management is the discipline ensuring that oceanographic data both past and present are available to researchers.

TRANSPORTATION ON THE OCEANS

For thousands of years, oceans provided one of the fastest and most valuable forms of transportation. By 3200 B.C.E., Egyptian ships made of reeds (tall, woody grass) used sails to travel along the coast of northern Africa. Over the centuries, ocean–going ships became larger and faster. Around 1000 B.C.E. the Vikings explored the coast of Canada in sailboats. Spanish ships explored the Americas in the fifteenth and sixteenth centuries. British tall ships carried settlers to the Americas, Asia, Australia, and Africa in the sixteenth through nineteenth centuries. Until the mid–

twentieth century, ships were the only mode of transportation for ocean crossings. The rise of air transportation after 1930 reduced the role of ocean-going vessels in transportation. Airplanes provided a quicker and often cheaper way to move people great distances, which caused the types of vessels and purposes of ocean transportation to change.

Immigration to the New World

For the first 450 years after the discovery of the New World, ships provided the only form of transportation between Europe and the Americas. Nearly every citizen of the United States is descended from ancestors who traveled to the New World by ship, and immigration to the New World was a major factor in ocean transportation during this time. Immigration patterns to the United States reflect that immigrants came from various countries in waves. The earliest settlers came from the British Isles and Africa. Before 1790, about 500,000 immigrants came to the United States from the British Isles, and 300,000 immigrants came from Africa.

The middle half of the nineteenth century saw a flood of immigrants from Europe with 3 million from the German Empire, 2.8 million from Ireland, and 2 million from England. The United States experienced its greatest influx of immigration between 1880 and 1930. During this period, nearly 20 million immigrants crossed the Atlantic Ocean on ships. These immigrants came primarily from Italy, Russia, Germany, Britain, and the Austro-Hungarian Empire. Twelve million of these immigrants entered the United States through Ellis Island, near New York City. Between 1897 and 1938, Ellis Island served as the main processing point for immigrants.

Today over 100 million Americans can trace their ancestry to an immigrant who landed on Ellis Island. Ocean transportation in America has a dark side. Slave ships transported tens of thousands of Africans to the New World every year. Between the sixteenth and nineteenth centuries, between 15 million and 20 million Africans were involuntarily brought to the Americas as slaves. About 400,000 slaves were transported to the British colonies and the United States. Scholars estimate that as many as 1 million African slaves died during ocean transit to the Americas.

Transatlantic journeys

Not all ocean crossing ships were only filled with immigrants. Travellers also used ships to cross the Atlantic Ocean to go between Europe and the Americas. In 1818, New York's Black Ball Line became the first company to offer regular travel across the Atlantic Ocean. The rise of steam ships in the mid-1800s made ocean crossings faster. While these ships focused on luxury travel for wealthy passengers, they also fueled immigration. Cruise liners offered low–cost, no frills transportation for many immigrants. The immigrants stayed in steerage class, the least expensive accommodations, and were often responsible for bringing their meals.

By the early twentieth century, cruise liner companies began to build larger and more luxurious ships, including Olympic, Lusitania, Britannic, and Titanic. These ships emphasized comfort and extravagance over speed. Many of these cruise liners contained swimming pools, dance halls, and tennis courts. Unfortunately, the superliners of the early nineteenth century did not stress safety. Thousands of lives were lost in the sinkings of the Titanic in 1912 and Lusitania in 1915.

The rise of the cruise ship

By 1950, airplanes replaced cruise liners as the main mode of transportation across the oceans. Many travellers did not choose to spend days crossing the ocean when it could be done in hours by plane. Cruise liner companies had to change their approach to fit the new reality of air travel. They could no longer market cruise liners as a form of transportation to take while on vacation. Instead, cruise companies began advertising cruise liners as a vacation by themselves. By focusing on exotic locales, such as the Caribbean and Mediterranean Seas, cruise companies found a willing audience.

In modern day cruise ships have swimming pools, cinemas, dance clubs, theatres, and classrooms. Modern cruise ships are subject to many safety regulations. Today nearly 8 million Americans go on cruises every year. Cruises generate about $18 billion every year for the United States' economy. A modern cruise ship carries about 2000 guests and 900 crew members. The largest cruise ship in the world as of 2004, Queen Mary 2, was 1,132 feet (345 meters) long and 151,400 gross tons (term describing the size of a boat, ship, or barge). Queen Mary 2 can carry 2,620 guests and 1,253 crew members. In 2004 Queen Mary 2 was the only passenger ship that made regular transatlantic journeys.

Ferries

Ferries are one of the most important forms of modern ocean transportation. Ferries are ships that carry people and, occasionally, cars over relative short distances. Some ferries are simple ships that transport only people. Ferries that transport people and cars are called "roll-on, roll-off" ships. Cars can quickly roll on these ferries upon departure and easily roll off upon arrival. While some ferries are simple boats, many ferries are technologically advanced ships, including hovercrafts or hydrofoils. A hovercraft is a ship that floats above the surface of the water on a cushion of air.

A rubber skirt is located between the main ship and the water. Air is pushed into the rubber skirt, creating a cushion of air. Hovercrafts offer smooth rides over rough seas. A hydrofoil is a ship that has wing-like foils (wing-like structures that raises part or all of a powerboat's hull out of the water) underneath the hull of the ship. As the boat increases speed, the foils lift the hull of the ship out of the water. Only the foils skim the top of the water. Like a hovercraft, the main body of a hydrofoil rides above the surface of the water. This reduces drag and increases speed.

Unlike most cruise ships, not all ferries are subject to strict safety regulations. Many passengers die in ferry accidents every year, mostly in the developing world. In 2002, the ferry Joola sank off the coast of Africa near Senegal. Joola was carrying over three times its capacity. Over 1,800 people died in the accident, which is more than the number of people who died on the Titanic.

OCEANIC CIRCULATION

A number of ocean currents flow with great persistence, setting up a circulation that continues with relatively little change throughout the year. Because of the influence of wind in creating current, there is a relationship between this oceanic circulation and the general circulation of the atmosphere. Some differences in opinion exist regarding the names and limits of some of the currents, but those shown are representative. Speed may vary somewhat with the season. This is particularly noticeable in the Indian Ocean and along the South China coast, where currents are influenced to a marked degree by the monsoons.

Southern Ocean Currents

The Southern Ocean has no meridional boundaries and its waters are free to circulate around the world. It serves as a conveyor belt for the other oceans, exchanging waters between them. The northern boundary of the Southern Ocean is marked by the Subtropical Convergence zone. This zone marks the transition from the temperate region of the ocean to the polar region and is associated with the surfacing of the main thermocline. This zone is typically found at 40°S but varies with longitude and season. In the Antarctic, the circulation is generally from west to east in a broad, slow-moving current extending completely around Antarctica. This is called the Antarctic Circumpolar Current or the West Wind Drift, and it is formed partly by the strong westerly wind in this area, and partly by density differences. This current is augmented by the Brazil and Falkland Currents in the Atlantic, the East Australia Current in the Pacific, and the Agulhas Current in the Indian Ocean.

In return, part of it curves northward to form the Cape Horn, Falkland, and most of the Benguela Currents in the Atlantic, and the Peru Current in the Pacific. In a narrow zone next to the Antarctic continent, a westward flowing coastal current is usually found. This current is called the East Wind Drift because it is attributed to the prevailing easterly winds which occur there.

Atlantic Ocean Currents

The trade winds set up a system of equatorial currents which at times extends over as much as 50° of latitude or more. There are two westerly flowing currents conforming generally with the areas of trade winds, separated by a weaker, easterly flowing countercurrent. The North Equatorial Current originates to the northward of the Cape Verde Islands and flows almost due west at an average speed of about 0.7 knot. The South Equatorial Current is more extensive. It starts off the west coast of Africa, south of the Gulf of Guinea, and flows in a generally westerly direction at an average speed of about 0.6 knot. However, the speed gradually increases until it may reach a value of 2.5 knots, or more, off the east coast of South America. As the current approaches Cabo de Sao Roque, the eastern extremity of South America, it divides, the southern part curving towards the south along the coast of Brazil, and the northern part being deflected northward by the continent of South America.

Between the North and South Equatorial Currents, the weaker North Equatorial Countercurrent sets towards the east in the general vicinity of the doldrums. This is fed by water from the two westerly flowing equatorial currents, particularly the South Equatorial Current. The extent and strength of the Equatorial Countercurrent changes with the seasonal variations of the wind. It reaches a maximum during July and August, when it extends from about 50° west longitude to the Gulf of Guinea. During its minimum, in December and January, it is of very limited extent, the western portion disappearing altogether.

That part of the South Equatorial Current flowing along the northern coast of South America which does not feed the Equatorial Countercurrent unites with the North Equatorial Current at a point west of the Equatorial Countercurrent. A large part of the combined current flows through various passages between the Windward Islands and into the Caribbean Sea. It sets towards the west, and then somewhat north of west, finally arriving off the Yucatan peninsula. From there, the water enters the Gulf of Mexico and forms the Loop Current; the path of the Loop Current is variable with a 13-month period.

It begins by flowing directly from Yucatan to the Florida Straits, but gradually grows to flow anticyclonically around the entire Eastern Gulf; it then collapses, again following the direct path from Yucatan to the Florida Straits, with the loop in the Eastern Gulf becoming a separate eddy which slowly flows into the Western Gulf.

Within the Straits of Florida, the Loop Current feeds the beginnings of the most remarkable of American ocean currents, the Gulf Stream. Off the southeast coast of Florida this current is augmented by the Antilles Current which flows along the northern coasts of Puerto Rico, Hispaniola, and Cuba. Another current flowing eastward of the Bahamas joins the stream north of these islands. The Gulf Stream follows generally along the east coast of North America, flowing around Florida, northward and then northeastward towards Cape Hatteras, and then curving towards the east and becoming broader and slower.

After passing the Grand Banks, it turns more towards the north and becomes a broad drift current flowing across the North Atlantic. The part in the Straits of Florida is sometimes called the Florida Current. A tremendous volume of water flows northward in the Gulf Stream. It can be distinguished by its deep indigo-blue colour, which contrasts sharply with the dull green of the surrounding water. It is accompanied by frequent squalls. When the Gulf Stream encounters the cold water of the Labrador Current, principally in the vicinity of the Grand Banks, there is little mixing of the waters.

Instead, the junction is marked by a sharp change in temperature. The line or surface along which this occurs is called the cold wall. When the warm Gulf Stream water encounters cold air, evaporation is so rapid that the rising vapour may be visible as frost smoke. Investigations have shown that the current itself is much narrower and faster than previously supposed, and considerably more variable in its position and speed. The maximum current off Florida ranges from about 2 to 4 knots. Northward, the speed is generally less, and it decreases further after the current passes Cape Hatteras.

As the stream meanders and shifts position, eddies sometimes break off and continue as separate, circular flows until they dissipate. Boats in the Newport-Bermuda sailing yacht race have been known to be within sight of each other and be carried in opposite directions by different parts of the same current. This race is generally won by the boat which catches an eddy just right. As the current shifts position, its extent does not always coincide with the area of warm, blue water. When the sea is relatively smooth, the edges of the current are marked by ripples. A recirculation region exists adjacent to and southwest of the Gulf Stream. The flow of water in the recirculation region is opposite to that in the Gulf Stream and surface currents are much weaker, generally less than half a knot. As the Gulf Stream continues eastward and northeastward beyond the Grand Banks, it gradually widens and decreases speed until it becomes a vast, slow-moving current known as the North Atlantic Current, in the general vicinity of the prevailing westerlies.

In the eastern part of the Atlantic it divides into the Northeast Drift Current and the Southeast Drift Current. The Northeast Drift Current continues in a generally northeasterly direction towards the Norwegian Sea. As it does so, it continues to widen and decrease speed. South of Iceland it branches to form the Irminger Current and the Norway Current. The Irminger Current curves towards the north and northwest to join the East Greenland Current southwest of Iceland.

The Norway Current continues in a northeasterly direction along the coast of Norway. Part of it, the North Cape Current, rounds North Cape into the Barents Sea. The other part curves towards the north and becomes known as the Spitsbergen Current. Before reaching Svalbard (Spitsbergen), it curves towards the west and joins the cold East Greenland Current flowing southward in the Greenland Sea.

As this current flows past Iceland, it is further augmented by the Irminger Current. Off Kap Farvel, at the southern tip of Greenland, the East Greenland Current curves sharply to the northwest following the coastline. As it does so, it becomes known as the West Greenland Current, and its character changes from that of an intense western boundary current to a weaker eastern boundary current. This current continues along the west coast of Greenland, through Davis Strait, and into Baffin Bay.

In Baffin Bay the West Greenland Current generally follows the coast, curving westward off Kap York to form the southerly flowing Labrador Current. This cold current flows southward off the coast of Baffin Island, through Davis Strait, along the coast of Labrador and Newfoundland, to the Grand Banks, carrying with it large quantities of ice. Here it encounters the warm water of the Gulf Stream, creating the cold wall. Some of the cold water flows southward along the east coast of North America, inshore of the Gulf Stream, as far as Cape Hatteras. The remainder curves towards the east and flows along the northern edge of the North Atlantic and Northeast Drift Currents, gradually merging with them. The Southeast Drift Current curves towards the east, southeast, and then south as it is deflected by the coast of Europe.

It flows past the Bay of Biscay, towards southeastern Europe and the Canary Islands, where it continues as the Canary Current. In the vicinity of the Cape Verde Islands, this current divides, part of it curving towards the west to help form the North Equatorial Current, and part of it curving towards the east to follow the coast of Africa into the Gulf of Guinea, where it is known as the Guinea Current. This current is augmented by the North Equatorial Countercurrent and, in summer, it is strengthened by monsoon winds. It flows in close proximity to the South Equatorial Current, but in the opposite direction.

As it curves towards the south, still following the African coast, it merges with the South Equatorial Current. The clockwise circulation of the North Atlantic leaves a large central area between the recirculation region and the Canary Current which has no well-defined currents. This area is known as the Sargasso Sea, from the large quantities of sargasso or gulfweed encountered there. That branch of the South Equatorial Current which curves towards the south off the east coast of South America, follows the coast as the warm, highly-saline Brazil Current, which in some respects resembles a weak Gulf Stream.

Off Uruguay it encounters the colder, less-salty Falkland or Malvinas Current forming a sharp meandering front in which eddies may form. The two currents curve towards the east to form the broad, slow-moving, South Atlantic Current in the general vicinity of the prevailing westerlies and the front dissipates somewhat. This current flows eastward to a point west of the Cape of Good Hope, where it curves northward to follow the west coast of Africa as the strong Benguela Current, augmented somewhat by part of the Agulhas Current flowing around the southern part of Africa from the Indian Ocean.

As it continues northward, the current gradually widens and slows. At a point east of St. Helena Island it curves westward to continue as part of the South Equatorial Current, thus completing the counterclockwise circulation of the South Atlantic. The Benguela Current is also augmented somewhat by the West Wind Drift, a current which flows easterly around Antarctica. As the West Wind Drift flows past Cape Horn, that part in the immediate vicinity of the cape is called the Cape Horn Current. This current rounds the cape and flows in a northerly and northeasterly direction along the coast of South America as the Falkland or Malvinas Current.

Pacific Ocean Currents

Pacific Ocean currents follow the general pattern of those in the Atlantic. The North Equatorial Current flows westward in the general area of the northeast trades, and the South Equatorial Current follows a similar path in the region of the southeast trades. Between these two, the weaker North Equatorial Countercurrent sets towards the east, just north of the equator. After passing the Mariana Islands, the major part of the North Equatorial Current curves somewhat towards the northwest, past the Philippines and Taiwan.

Here it is deflected further towards the north, where it becomes known as the Kuroshio, and then towards the northeast past the Nansei Shoto and Japan, and on in a more easterly direction. Part of the Kuroshio, called the Tsushima Current, flows through Tsushima Strait, between Japan and Korea, and the Sea of Japan, following generally the northwest coast of Japan. North of Japan it curves eastward and then southeastward to rejoin the main part of the Kuroshio. The limits and volume of the Kuroshio are influenced by the monsoons, being augmented during the season of southwesterly winds, and diminished when the northeasterly winds are prevalent. The Kuroshio (Japanese for "Black Stream") is so named because of the dark colour of its water. It is sometimes called the Japan Current. In many respects it is similar to the Gulf Stream of the Atlantic.

Like that current, it carries large quantities of warm tropical water to higher latitudes, and then curves towards the east as a major part of the general clockwise circulation in the Northern Hemisphere. As it does so, it widens and slows, continuing on between the Aleutians and the Hawaiian Islands, where it becomes known as the North Pacific Current. As this current approaches the North American continent, most of it is deflected towards the right to form a clockwise circulation between the west coast of North America and the Hawaiian Islands called the California Current. This part of the current has become so broad that the circulation is generally weak.

Near the coast, the southeastward flow intensifies and average speeds are about 0.8 knot. But the flow pattern is complex, with offshore directed jets often found near more prominent capes, and poleward flow often found over the upper slope and outer continental shelf. It is strongest near land. Near the southern end of Baja California, this current curves sharply to the west and broadens to form the major portion of the North Equatorial Current. During the winter, a weak countercurrent flows northwestward, inshore of the southeastward flowing California Current, along the west coast of North America from Baja California to Vancouver Island.

This is called the Davidson Current. Off the west coast of Mexico, south of Baja California the current flows southeastward during the winter as a continuation of part of the California Current. During the summer, the current in this area is northwestward as a continuation of the North Equatorial Countercurrent. As in the Atlantic, there is in the Pacific a counterclockwise circulation to the north of the clockwise circulation.

Cold water flowing southward through the western part of Bering Strait between Alaska and Siberia, is joined by water circulating counterclockwise in the Bering Sea to form the Oyashio. As the current leaves the strait, it curves towards the right and flows southwesterly along the coast of Siberia and the Kuril Islands. This current brings quantities of sea ice, but no icebergs. When it encounters the Kuroshio, the Oyashio curves southward and then eastward, the greater portion joining the Kuroshio and North Pacific Current. The northern branch of the North Pacific Current curves in a counterclockwise direction to form the Alaska Current, which generally follows the coast of Canada and Alaska. When the Alaska Current turns to the southwest and flows along the Kodiak Island and the Alaska Peninsula, its character changes to that of a western boundary current and it is called the Alaska Stream. When this westward flow arrives off the Aleutian Islands, it is less intense and becomes known as the Aleutian Current. Part of it flows along the southern side of these islands to about the 180th meridian, where it curves in a counter-clockwise direction and becomes an easterly flowing current, being augmented by the northern part of the Oyashio.

The other part of the Aleutian Current flows through various openings between the Aleutian Islands, into the Bering Sea. Here it flows in a general counterclockwise direction. The southward flow along the Kamchatka peninsula is called the Kamchatka Current which feeds the southerly flowing Oyashio. Some water flows northward from the Bering Sea through the eastern side of the Bering Strait, into the Arctic Ocean. The South Equatorial Current, extending in width between about 4°N latitude and 10°S, flows westward from South America to the western Pacific.

After this current crosses the 180th meridian, the major part curves in a counterclockwise direction, entering the Coral Sea, and then curving more sharply towards the south along the east coast of Australia, where it is known as the East Australian Current. The East Australian Current is the weakest of the subtropical western boundary currents and separates from the Australian coast near 34°S. The path of the current from Australia to New Zealand is known as the Tasman Front, which marks the boundary between the warm water of the Coral Sea and the colder water of the Tasman Sea. The continuation of the East Australian Current east of New Zealand is the East Auckland Current. The East Auckland Current varies seasonally: in winter, it separates from the shelf and flows eastward, merging with the West Wind Drift, while in winter it follows the New Zealand shelf southward as the East Cape Current until it reaches Chatham Rise where it turns eastward, thence merging with the West Wind Drift. Near the southern extremity of South America, most of this current flows eastward into the Atlantic, but part of it curves towards the left and flows generally northward along the west coast of South America as the Peru Current or Humboldt Current. Occasionally a set directly towards land is encountered.

At about Cabo Blanco, where the coast falls away to the right, the current curves towards the left, past the Galapagos Islands, where it takes a westerly set and constitutes the major portion of the South Equatorial Current, thus completing the counterclockwise circulation of the South Pacific. During the northern hemisphere summer, a weak northern branch of the South Equatorial Current, known as the New Guinea Coastal Current, continues on towards the west and northwest along both the southern and northeastern coasts of New Guinea. The southern part flows through Torres Strait, between New Guinea and Australia, into the Arafura Sea. Here, it gradually loses its identity, part of it flowing on towards the west as part of the South Equatorial Current of the Indian Ocean, and part of it following the coast of Australia and finally joining the easterly flowing West Wind Drift. The northern part of New Guinea Coastal Current both curves in a clockwise direction to help form the Pacific Equatorial Countercurrent and off Mindanao turns southward to form a southward flowing boundary current called the Mindanao Current. During the northern hemisphere winter, the New Guinea Coastal Current may reverse direction for a few months.

Indian Ocean Currents

Indian Ocean currents follow generally the pattern of the Atlantic and Pacific but with differences caused principally by the monsoons, the more limited extent of water in the Northern Hemisphere, and by limited communication with the Pacific Ocean along the eastern boundary. During the northern hemisphere winter, the North Equatorial Current and South Equatorial Current flow towards the west, with the weaker, eastward Equatorial Countercurrent flowing between them, as in the Atlantic and Pacific (but somewhat south of the equator). But during the northern hemisphere summer, both the North Equatorial Current and the Equatorial Counter current are replaced by the Southwest Monsoon Current, which flows eastward and southeastward across the Arabian Sea and the Bay of Bengal.

Near Sumatra, this current curves in a clockwise direction and flows westward, augmenting the South Equatorial Current, and setting up a clockwise circulation in the northern part of the Indian Ocean. Off the coast of Somalia, the Somali Current reverses direction during the northern hemisphere summer with northward currents reaching speeds of 5 knots or more. Twice a year, around May and November, westerly winds along the equator result in an eastward Equatorial Jet which feeds warm water towards Sumatra. As the South Equatorial Current approaches the coast of Africa, it curves towards the southwest, part of it flowing through the Mozambique Channel between Madagascar and the mainland, and part flowing along the east coast of Madagascar. At the southern end of this island the two join to form the strong Agulhas Current, which is analogous to the Gulf Stream.

This current, when opposed by strong winds from Southern Ocean storms, creates dangerously large seas. South of South Africa, the Agulhas Current retroflects, and most of the flow curves sharply southward and then eastward to join the West Wind Drift; this junction is often marked by a broken and confused sea, made much worse by westerly storms. A small part of the Agulhas Current rounds the southern end of Africa and helps form the Benguela Current; occasionally, strong eddies are formed in the retroflection region and these too move into the Southeastern Atlantic. The eastern boundary currents in the Indian Ocean are quite different from those found in the Atlantic and Pacific. The seasonally reversing South Java Current has strongest westward flow during August when monsoon winds are easterly and the Equatorial jet is inactive. Along the coast of Australia, a vigourous poleward flow, the Leeuwin Current, runs against the prevailing winds.

OBSERVATIONS OF THE DEEP CIRCULATION

The abyssal circulation is less well known than the upper-ocean circulation. Direct observations from moored current meters or deep-drifting floats were difficult to make until recently, and there are few long-term direct measurements of current. In addition, the measurements do not produce a stable mean value for the deep currents. For example, if the deep circulation takes roughly 1,000 years to transport water from the north Atlantic to the Antarctic Circumpolar Current and then to the north

Pacific, the mean flow is about 1mm/s. Observing this small mean flow in the presence of typical deep currents having variable velocities of up to 10 cm/s or greater, is very difficult. Most of our knowledge of the deep circulation is inferred from measured distribution of temperature, salinity, oxygen, silicate, tritium, fluorocarbons and other tracers. These measurements are much more stable than direct current measurements, and observations made decades apart can be used to trace the circulation. Tomczak carefully describes how the techniques can be made quantitative and how they can be applied in practice.

Water Masses

The concept of water masses originates in meteorology. Vilhelm Bjerknes, a Norwegian meteorologist, first described the cold air masses that form in the polar regions. He showed how they move southward, where they collide with warm air masses at places he called fronts, just as masses of troops collide at fronts in war. In a similar way, water masses are formed in different regions of the ocean, and the water masses are often separated by fronts. Note, however, that strong winds are associated with fronts in the atmosphere because of the large difference in density and temperature on either side of the front. Fronts in the ocean sometimes have little contrast in density, and these fronts have only weak currents.

Tomczak defines a water mass as a body of water with a common formation history, having its origin in a physical region of the ocean. Just as air masses in the atmosphere, water masses are physical entities with a measurable volume and therefore occupy a finite volume in the ocean. In their formation region they have exclusive occupation of a particular part of the ocean. Elsewhere they share the ocean with other water masses with which they mix. The total volume of a water mass is given by the sum of all its elements regardless of their location.

Plots of salinity as a function of temperature, called T-S plots, are used to delineate water masses and their geographical distribution, to describe mixing among water masses, and to infer motion of water in the deep ocean. Here's why the plots are so useful: water properties, such as temperature and salinity, are formed only when the water is at the surface or in the mixed layer. Heating, cooling, rain, and evaporation all contribute. Once the water sinks below the mixed layer, temperature and salinity can change only by mixing with adjacent water masses. Thus water from a particular region has a particular temperature associated with a particular salinity, and the relationship changes little as the water moves through the deep ocean.

DEEP CIRCULATION IN THE OCEAN

The direct forcing of the oceanic circulation by wind discussed in the last few stages is limited mostly to the upper kilometer of the water column. Below a kilometer lie the vast water masses of the ocean extending to depths of 4 km –5 km. The water is everywhere cold, with a potential temperature less than 4°C. The water mass is formed when cold, dense water sinks from the surface to great depths at high latitudes. It spreads out from these regions to fill the ocean basins. Deep mixing eventually pulls the water up through the thermocline over large areas of the ocean. It is this upwelling that drives the deep circulation. The vast deep ocean is usually referred to as the abyss, and the circulation as the abyssal circulation.

The densest water at the sea surface, water that is dense enough to sink to the bottom, is formed when frigid air blows across the ocean at high latitudes in winter in the Atlantic between Norway and Greenland and near Antarctica. The wind cools and evaporates water. If the wind is cold enough, sea ice forms, further increasing the salinity of the water because ice is fresher than sea water. Bottom water is produced only in these two regions. In other polar regions, cold, dense water is formed, but it is not quite salty enough to sink to the bottom.

At mid and low latitudes, the density, even in winter, is sufficiently low that the water cannot sink more than a few hundred meters into the ocean. The only exception are some seas, such as the Mediterranean Sea, where evaporation is so great that the salinity of the water is sufficiently great for the water to sink to intermediate depths in the seas. If these seas are can exchange water with the open ocean, the waters formed in winter in the seas spreads out to intermediate depths in the ocean.

Defining the Deep Circulation

Many terms have been used to describe the deep circulation.

They include:

- Abyssal circulation;
- Thermohaline circulation;
- Meridional overturning circulation; and
- Global conveyor.

The term thermohaline circulation was once widely used, but it has disappeared almost entirely from the oceanographic literature. It is no longer used because it is now clear that the flow is not density driven, and because the concept has not been clearly defined. The meridional overturning circulation is better defined. It is the zonal average of the flow plotted as a function of depth and latitude. Plots of the circulation show where vertical flow is important, but they show no information about how circulation in the gyres influences the flow.

Following Wunsch, I define the deep circulation as the circulation of mass. Of course, the mass circulation also carries heat, salt, oxygen, and other properties. But the circulation of the other properties is not the same as the mass transport. For example, Wunsch points out that the North Atlantic imports heat but exports oxygen. The deep circulation is mostly wind driven, but tidal mixing is also important. The wind enters several ways. It cools the surface and evaporates water, which determines where deep convection occurs. And, it produces turbulence in the deep ocean which mixes cold water upward.

IMPORTANCE OF THE DEEP CIRCULATION

The deep circulation which carries cold water from high latitudes in winter to lower latitudes throughout the world has very important consequences.

- The contrast between the cold deep water and the warm surface waters determines the stratification of the oceans. Stratification strongly influences ocean dynamics.
- The volume of deep water is far larger than the volume of surface water. Although currents in the deep ocean are relatively weak, they have transports comparable to the surface transports.
- The fluxes of heat and other variables carried by the deep circulation influences Earth's heat budget and climate. The fluxes vary from decades to centuries to millennia, and this variability is thought to modulate climate over such time intervals. The ocean may be the primary cause of variability over times ranging from years to decades, and it may have helped modulate ice-age climate.

Two aspects of the deep circulation are especially important for understanding Earth's climate and its possible response to increased carbon dioxide CO_2 in the atmosphere:

1. The ability of cold water to absorb CO2 from the atmosphere, and
2. The ability of deep currents to modulate the heat transported from the tropics to high latitudes.

The Oceans as a Reservoir of Carbon Dioxide

The oceans are the primary reservoir of readily available CO_2, an important greenhouse gas. The oceans contain 40,000 GtC of dissolved, particulate, and living forms of carbon. The land contains 2,200 GtC, and the atmosphere contains only 750 GtC. Thus the oceans hold 50 times more carbon than the air. Furthermore, the amount of new carbon put into the atmosphere since the industrial revolution, 150 GtC, is less than the amount of carbon cycled through the marine ecosystem in five years. (1 GtC = 1 gigaton of carbon = 1012 kilograms of carbon.) Carbonate rocks such as limestone, the shells of marine animals, and coral are other, much larger, reservoirs. But this carbon is locked up. It cannot be easily exchanged with carbon in other reservoirs.

More CO_2 dissolves in cold water than in warm water. Just imagine shaking and opening a hot can of CokeTM. The CO_2 from a hot can will spew out far faster than from a cold can. Thus the cold deep water in the ocean is the major reservoir of dissolved CO_2 in the ocean.

New CO_2 is released into the atmosphere when fossil fuels and trees are burned. Very quickly, 48% of the CO_2 released into the atmosphere dissolves in the cold waters of the ocean, much of which ends up deep in the ocean.

Forecasts of future climate change depend strongly on how much CO_2 is stored in the ocean and for how long. If little is stored, or if it is stored and later released into the atmosphere, the concentration in the atmosphere will change, modulating Earth's long-wave radiation balance. How much and how long CO_2 is stored in the ocean depends on the deep circulation and the net flux of carbon deposited on the seafloor. The amount that dissolves depends on the temperature of the deep water, the storage time in the deep ocean depends on the rate at which deep water is replenished, and the deposition depends on whether the dead plants and animals that drop to the sea floor are oxidized. Increased ventilation of deep layers, and warming of the deep layers could release large quantities of the gas to the atmosphere.

The storage of carbon in the ocean also depends on the dynamics of marine ecosystems, upwelling, and the amount of dead plants and animals stored in sediments. But we won't consider these processes.

BIOLOGICAL OCEANOGRAPHY

Biological oceanographers (or marine biologists) focus on the patterns and distribution of marine organisms. These scientists work to understand why certain animals, plants, and microorganisms are found in different places and how these organisms grow. A variety of factors influence the success of a certain species in any location, including the chemistry and physical properties of the water. In turn, the biological organisms in the ocean affect the oceans on a global and local level. Biological oceanographers study all types of organisms that live in the ocean, from the very small to the very large.

They investigate patterns and distributions of the microscopic organisms including viruses (which are not really organisms, but genetic material such as DNA that do have the ability to reproduce), bacteria, and plankton (free-floating animals and plants). They also study the larger animals and plants, like kelp, seaweed, marine invertebrates (animals without a backbone), fish, and marine mammals. They incorporate information and techniques from a broad range of disciplines including chemistry, physics, remote sensing (the use of specialized instruments, such as satellites, to relay information about one location to another location for analysis), paleontology (study of fossils), and geography (study of Earth's surface) for their research.

Chemical oceanography

Chemical oceanographers study the chemicals that are dissolved in the ocean waters. Different parts of the ocean contain varying concentrations of gasses, salts, and other chemical components. These variations are due to the impact of the atmosphere, surrounding lands, seafloor, and biological organisms in the ocean water. Chemical oceanographers work to develop theories that explain the various patterns throughout the oceans. One of the more important problems facing chemical oceanographers today is understanding the concentration of and changes in carbon dioxide in the ocean. Carbon dioxide is a major greenhouse gas, meaning it holds a lot of heat when it is found as a gas in the atmosphere. Burning fossil fuels for industry and in cars releases carbon dioxide into the atmosphere, where it contributes to global warming. The ocean, however, can remove a lot of carbon dioxide from the atmosphere. Carbon dioxide readily combines with seawater. It then goes through a series of complex chemical reactions before it becomes a solid material called calcium carbonate. Calcium carbonate can be buried in the sediments (particles of gravel, sand, and clay) at the bottom of the ocean. This means that the ocean has the potential to act as a "sink" for a lot of the carbon dioxide in the atmosphere. Chemical oceanographers are working to determine just how large the sink is and how quickly it can act.

Physical oceanography

Physical oceanographers study the physical properties of the ocean. These include temperature, salinity, density, and ability to transmit light and sound. In turn, these fundamental physical characteristics affect the way that ocean currents move, the forces associated with waves, and the amount of energy absorbed by the ocean. The temperature and salinity of the water affect the density of the water. Cooler and saltier water sinks while warmer and fresher water floats. This seemingly simple property of the ocean drives much of the water circulation throughout the globe.

Density also affects the way that sound travels through water and the buoyancy (ability to float) of marine organisms. Some of the projects that physical oceanographers are studying include understanding trends in climate. Satellites measure ocean temperatures over the whole globe to try to discriminate between local changes in ocean temperature, like the El Niño-La Niña, a cycle that brings warm water and storms to the Eastern Pacific every 5 to 7 years, from more large scale changes, like global warming.

7

Geomorphology and Tectonics

Another basic assumption involves climatic morphogenesis, emphasizing the role of climatically controlled processes of landform genesis. Several of these concepts have yielded major intellectual controversy, such as the role of cataclysmic processes in shaping the landscape. These concepts apply to geomorphology of all scales.

APPLIED GEOMORPHOLOGY

Geomorphology has traditionally focused on the study of landforms and on the processes involved in their formation. Applied geomorphology is the practical application of this study to a range of environmental issues, both in terms of current problems and of future prediction. Applied geomorphology provides a strategic tool for informed decision-making in public policy development and in environmental resource management. Key areas of application include specific environmental settings, such as the coastal zone or dryland environments; the impacts of land use and management practice on Earth surface processes; and areas susceptible to natural hazards.

Over 60 per cent of the world's population live in the coastal zone in environments ranging from coral atolls, reclaimed or natural wetlands, dune-backed beaches, and barrier islands to cliff tops. Settlements under threat from coastal erosion and flooding from storm events, sea surges, and rising sea level lobby for protective engineering measures to prevent loss of property, livelihood, and life. Geomorphology has several applications in settings of this type.

An understanding of coastal landforms and the processes acting upon them can be used to map areas at risk from cliff failure, beach erosion, and flooding. This approach is of interest to potential developers and the insurance industry and is an important tool in environmental impact assessment. An understanding of the geomorphology of the coastal zone can also be used to predict the effects of modifying the coastal system. The installation of groynes, breakwaters, or protective sea walls has knock-on effects on the natural circulation of water and sediment in the near-shore environment. Artificially stabilizing cliffs to prevent erosion may seem the obvious solution for cliff-top dwellers, but a geomorphological evaluation might predict that this approach could starve beaches of the sediment provided by natural cliff fall, with a consequent impact on longshore drift of sediment, and would relocate the focus of erosion further along the coast.

The nature of the problem may thus change from cliff failure at one site to beach erosion and subsequent flooding at another. An understanding of the nature and complexity of coastal dynamics is thus an essential component of a coastal-zone management strategy and is important in predicting the future effects on coastal landforms of a rise in sea level. River-management strategies for flood alleviation have often adopted engineering solutions concentrated in particular river reaches, which are usually in areas of urban development.

Reach-specific intervention measures include lining the natural channel with concrete to prevent erosion and bank instability, channel straightening to force flood water to flow rapidly through particular reaches, and flow-control structures such as sluice gates and reservoirs to control water level. These artificial measures are not always successful in preventing flooding and erosion within the river catchment, and natural sections further downstream may be overwhelmed by the river at peak flood.

The engineered reaches of rivers often become a sterile landscape because fast-flowing water in a concrete-lined channel, with minimal variation in water depth and channel cross-section, provides a poor habitat for wetland flora and fauna. Geomorphology has been applied to 'river restoration' to recreate an integrated river management strategy within artificially created river systems, maximizing biodiversity while controlling river-flow conditions.

Applied geomorphology uses a holistic approach to river response at a catchment-wide scale; the basis here is an understanding of the relationships between river form and process, sediment transport, and the important role of river-bank vegetation. Certain landscapes have specific properties that impinge on our use and development of the environment. In cold environments, the presence of ground ice leads to problems in construction, communication, and housing. In permafrost zones, the ground is permanently frozen except for the upper layers of the soil, which thaw in the summer.

The upper soil, known as the active layer, is subject to repetitive cycles of freezing and thawing, making it geomorphologically active. The ground within the active layer will suffer heaving and deformation, disrupting communications and making road construction impracticable. Applied geomorphology can be used in mapping the active layer and ground ice in areas with differing rocks and sediments.

This information is then used to evaluate the problems that are likely to affect these areas. Ground heaving depends on the depth of the active layer and the type of sediment present; fine-grained silts present more of a problem than gravels. Additional problems in permafrost areas, as, for example, in some regions of Alaska, occur where structures have suffered dramatic subsidence as a result of heating in buildings.

Without appropriate insulation, heat radiates downwards from the building into the ground, thaws the underlying ice, and increases the depth of the active layer, thus effectively changing the structure of the soil. Applied geomorphology is consequently essential in land-use planning and site evaluation, in order to recognize such potential problems as land subsidence, slope instability, invasion of windblown sand, and impacts on natural drainage systems. Land used for agricultural production may suffer from degradation and desertification as a result of soil erosion, landsliding, and over-extraction of water for irrigation. Much agricultural practice focuses on maximizing yield and profit, often using techniques that can be detrimental to the environment, both in the short and the long term.

Applied geomorphology uses an understanding of the relationships between surface conditions, climate, vegetation, and soil erosion to advise farmers and politicians on how to improve land management for sustainable use of land and water resources. Natural hazards such as volcanic eruptions, earthquakes, and mudflows present a significant risk to the population of the surrounding area. Geomorphological mapping can be used to assess the present condition of the landscape and provide a hazard map.

The expression of a disaster may result in one settlement having significantly different risk assessment. For example, a volcanic eruption may pose a threat from volcanic ash and lava flows, pyroclastic flows, and bombardment from superheated volcanic bombs or associated hazards such as mudflows, depending on topography, soil cover, type of eruption, and predominant wind direction.

This application of geomorphological analysis is of significant interest to the emergency services and the insurance industry. Applied geomorphology can be used in modelling change to landforms and surface processes. This can include change from human impact on the environment to future prediction of climate change, from short-term El Niño and tropical storm events to longer-term change resulting from greenhouse warming and rising sea levels. In this way, applied geomorphology has a key role in managing the environment to minimize potential degradation of land, water, and natural resources.

GEOMORPHOLOGY-REAL-LIFE APPLICATIONS

Subsidence

Subsidence refers to the process of subsiding (settling or descending), on the part of either an air column or the solid earth, or, in the case of solid earth, to the resulting formation or depression. Subsidence in the atmosphere is discussed briefly in the entry Convection. Subsidence that occurs in the solid earth, known as geologic subsidence, is the settling or sinking by a body of rock or sediment. (The latter can be defined as material deposited at or near Earth's surface from a number of sources, most notably preexisting rock.)

As noted earlier, many geomorphologic processes can be caused either by nature or by human beings. An example of natural subsidence takes place in the aftermath of an earthquake, during which large areas of solid earth may simply drop by several feet. Another example can be observed at the top of a volcano some time after it has erupted, when it has expelled much of its material (i.e., magma) and, as a result, has collapsed.

Natural subsidence also may result from cave formation in places where underground water has worn away limestone. If the water erodes too much limestone, the ceiling of the cave will subside, usually forming a sinkhole at the surface.

The sinkhole may fill with water, making a lake; the formation of such sinkholes in many spots throughout an area (whether the sinkholes become lakes or not), is known as karst topography.

In places where the bedrock is limestone—particularly in the sedimentary basins of rivers—karst topography is likely to develop. The United States contains the most extensive karst region in the world, including the Mammoth cave system in Kentucky. Karst topography is very pronounced in the hills of southern China, and karst landscapes have been a prominent feature of Chinese art for centuries. Other extensive karst regions can be found in southern France, Central America, Turkey, Ireland, and England.

Man-made Subsidence

Man-made subsidence often ensues from the removal of groundwater or fossil fuels, such as petroleum or coal. Groundwater removal can be perfectly safe, assuming the area experiences sufficient rainfall to replace, or recharge, the lost water. If recharging does not occur in the necessary proportions, however, the result will be the eventual collapse of the aquifer, a layer of rock that holds groundwater.

In so-called room-and-pillar coal mining, pillars, or vertical columns, of coal are left standing, while the areas around them are extracted. This method maintains the ceiling of the "room" that has been mined of its coal. After the mine is abandoned, however, the pillar eventually may experience so much stress that it breaks, leading to the collapse of the mined room. As when the ceiling of a cave collapses, the subsidence of a coal mine leaves a visible depression above ground.

Uplift

As its name implies, uplift describes a process and results opposite to those of subsidence. In uplift the surface of Earth rises, owing either to a decrease in downward force or to an increase in upward force. One of the most prominent examples of uplift is seen when plates collide, as when India careened into the southern edge of the Eurasian landmass some 55 million years ago. The result has been a string of mountain ranges, including the Himalayas, Karakoram Range, and Hindu Kush, that contain most of the world's tallest peaks.

Plates move at exceedingly slow speeds, but their mass is enormous. This means that their inertia (the tendency of a moving object to keep moving unless acted upon by an outside force) is likewise gargantuan in scale.

Therefore, when plates collide, though they are moving at a rate equal to only a few inches a year, they will keep pushing into each other like two automobiles crumpling in a head-on collision. Whereas a car crash is over in a matter of seconds, however, the crumpling of continental masses takes place over hundreds of thousands of years.

When sea floor collides with sea floor, one of the plates likely will be pushed under by the other one, and, likewise, when sea floor collides with continental crust, the latter will push the sea floor under. This results in the formation of volcanic mountains, such as the Andes of South America or the Cascades of the Pacific Northwest, or volcanic islands, such as those of Japan, Indonesia, or Alaska's Aleutian chain.

Isostatic Compensation

In many other instances, collision, compression, and extension cause uplift. On the other hand, as noted, uplift may result from the removal of a weight. This occurs at the end of an ice age, when glaciers as thick as 1.9 mi. (3 km) melt, gradually removing a vast weight pressing down on the surface below.

This movement leads to what is called isostatic compensation, or isostatic rebound, as the crust pushes upward like a seat cushion rising after a person is longer sitting on it. Scandinavia is still experiencing uplift at a rate of about 0.5 in. (1 cm) per year as the after-effect of glacial melting from the last ice age. The latter ended some 10,000 years ago, but in geologic terms this is equivalent to a few minutes' time on the human scale.

Islands

Geomorphology, as noted earlier, is concerned with landforms, such as mountains and volcanoes as well as larger ones, including islands and even continents. Islands present a particularly interesting area of geomorphologic study. In general, islands have certain specific characteristics in terms of their land structure and can be analysed from the standpoint of the geosphere, but particular islands also have unique ecosystems, requiring an interdisciplinary study that draws on botany, zoology, and other subjects.

In addition, there is something about an island that has always appealed to the human imagination, as evidenced by the many myths, legends, and stories about islands.

Some examples include Homer's *Odyssey,* in which the hero Odysseus visits various islands in his long wanderings; Thomas More's *Utopia,* describing an idealized island republic; *Robinson Crusoe,* by Daniel Defoe, in which the eponymous hero lives for many years on an island with no companion but the trusty native Friday; *Treasure Island,* by Robert Louis Stevenson, in which the island is the focus of a treasure hunt; and Mark Twain's *Adventures of Huckleberry Finn,* depicting Jackson Island in the Mississippi River, to which Huckleberry Finn flees to escape "civilization."

One of the favourite subjects of cartoonists is that of a castaway stranded on a desert island, a mound of sand with no more than a single tree. Movies, too, have long portrayed scenarios, from the idyllic to the brutal, that take place on islands, particularly deserted ones, a notable example being *Cast Away* (2000). A famous line by the English poet John Donne (1572-1631) warns that "no man is an island," implying that many wish they could enjoy the independence suggested by the concept of an island. Within the Earth system, however, nothing is fully independent, and, as we shall see, this is certainly the case where islands are concerned.

The Islands of Earth

Earth has literally tens of thousands of islands. Just two archipelagos (island chains), those that make up the Philippines and Indonesia, include thousands of islands each. While there are just a few dozen notable islands on Earth, many more dot the planet's seas and oceans. The largest are these:

- Greenland (Danish, northern Atlantic): 839,999 sq. mi.(2,175,597 sq km)
- New Guinea (divided between Indonesia and Papua New Guinea, western Pacific): 316,615 sq. mi. (820,033 sq km)
- Borneo (divided between Indonesia and Malaysia, western Pacific): 286,914 sq. mi. (743,107 sq km)
- Madagascar (Malagasy Republic, western Indian Ocean): 226,657 sq. mi. (587,042 sq km)
- Baffin (Canadian, northern Atlantic): 183,810 sq. mi. (476,068 sq km)
- Sumatra (Indonesian, northeastern Indian Ocean): 182,859 sq. mi. (473,605 sq km).

The list could go on and on, but it stops at Sumatra because the next-largest island, Honshu (part of Japan), is less than half as large, at 88,925 sq. mi. (230,316 sq km). Clearly, not all islands are created equal, and though some are heavily populated or enjoy the status of independent nations (e.g., Great Britain at number eight or Cuba at number 15), they are not necessarily the largest. On the other hand, some of the largest are among the most sparsely populated.

Of the 32 largest islands in the world, more than a third are in the icy northern Atlantic and Arctic, with populations that are small or practically nonexistent. Greenland's population, for instance, was just over 59,000 in 1998, while that of Baffin Island was about 13,200. On both islands, then, each person has about 14 frozen sq. mi. (22 sq km) to himself or herself, making.

Continents, Oceans, and Islands

Australia, of course, is not an island but a continent, a difference that is not related directly to size. If Australia *were* an island, it would be by far the largest. Australia is regarded as a continent, however, because it is one of the principal landmasses of the Indo-Australian plate, which is among a handful of major continental plates on Earth.

Whereas continents are more or less permanent (though they have experienced considerable rearrangement over the eons), islands come and go, seldom lasting more than 10 million years. Erosion or rising sea levels remove islands, while volcanic explosions can create new ones, as when an eruption off the coast of Iceland resulted in the formation of an island, Surtsey, in 1963.

Islands are of two types, continental and oceanic. Continental islands are part of continental shelves (the submerged, sloping ledges of continents) and may be formed in one of two ways. Rising ocean waters either cover a coastal region, leaving only the tallest mountains exposed as islands or cut off part of a peninsula, which then becomes an island.

Most of Earth's significant islands are continental and are easily spotted as such, because they lie at close proximity to continental landmasses. Many other continental islands are very small, however; examples include the barrier islands that line the East Coast of the United States. Formed from mainland sand brought to the coast by rivers, these are technically not continental islands, but they more clearly fit into that category than into the grouping of oceanic islands.

Oceanic islands, of which the Hawaiian-Emperor island chain and the Aleutians off the Alaskan coast are examples, form as a result of volcanic activity on the ocean floor. In most cases, there is a region of high volcanic activity, called a hot spot, beneath the plates, which move across the hot spot. This is the situation in Hawaii, and it explains why the volcanoes on the southern islands are still active while those to the north are not: the islands themselves are moving north across the hot spot. If two plates converge and one subducts, a deep trench with a parallel chain of volcanic islands may develop. Exemplified by the Aleutians, these chains are called island arcs.

Island Ecosystems

The ecosystem, or community of all living organisms, on islands can be unique owing to their separation from continents. The number of life-forms on an island is relatively small and can encompass some unusual circumstances compared with the larger ecosystems of continents. Ireland, for instance, has no native snakes, a fact "explained" by the legend that Saint Patrick drove them away. Hawaii and Iceland are also blessedly free of serpents.

Oceanic islands, of course, tend to have more unique ecosystems than do continental islands. The number of land-based animal life-forms is necessarily small, whereas the varieties of birds, flying insects, and surrounding marine life will be greater owing to those creatures' mobility across water. Vegetation is relatively varied, given the fact that winds, water currents, and birds may carry seeds.

Nonetheless, ecosystems of islands tend to be fairly delicate and can be upset by the human introduction of new predators (e.g., dogs) or new creatures to consume plant life (e.g., sheep). These changes sometimes can have disastrous effects on the overall balance of life on islands. Overgrazing may even open up the possibility of erosion, which has the potential of bringing an end to an island's life.

STRUCTURE AND TECTONICS

Drainage may adjust passively to varying resistance of geologic materials, or it may be actively induced to follow a particular course by tectonism. Examples of the latter include faulting, as in the Ganges-Brahmaputra delta region. Growing folds and domes have affected drainage in the Colorado Plateau and central Australia. Subsidence has been important in the Mississippi and Pantanal regions.

Streams that emerge from mountain fronts onto surrounding plains display a fascinating array of structural and tectonic controls. Where mountain fronts are erosional because of a complex interplay of geomorphic variables, they may develop flanking surfaces of planation called pediments. Deposition at the mountain front produces alluvial fans because of the tremendous increase in width as a stream emerges from a mountain canyon. Examples include the Tian Shan, Kosi, and Pantanal areas. Passive adjustment to structure is a quality of nearly all the study areas.

Perhaps the most interesting situations, however, are drainage anomalies, where streams cut across structural zones. Some streams appear to take the most difficult routes possible through fold belts. In his studies of the Appalachians and the Zagros Mountains, Oberlander has applied the term "obstinate streams" to this phenomenon. The Finke River is an excellent example. The Colorado River provides other examples.

Channel Patterns

Rivers display a remarkable variety of channel patterns that are especially amenable to study using spaceborne remote sensing systems.

The patterns relate to large-scale conditions of climate and tectonism that can only be appreciated on a global perspective. It is remarkable that, despite the geologic dominance of "big rivers", it is precisely those rivers that have received the least study. Experimental work by Schumm has done much to increase our understanding of channel patterns. Pattern adjustments, measured as sinuosity variation, are closely related to the type, size, and amount of sediment load.

They are also related to bank resistance and to the discharge characteristics of the stream. Many of the morphological dependencies of river patterns can be summarized in the following expressions:

$$Q_w \mu \frac{W, d, l}{S}$$

$$Q_s \mu \frac{W, l, S}{d, P}$$

These relationships are expressed by a large number of empirical equations treating the important independent variables, Q_w, a measure of mean annual water discharge, and Q_s, a measure of the type of sediment load. The dependent variables are the channel width, W, depth, d, the slope of the river channel, S, the sinuosity, P and the meander wavelength.

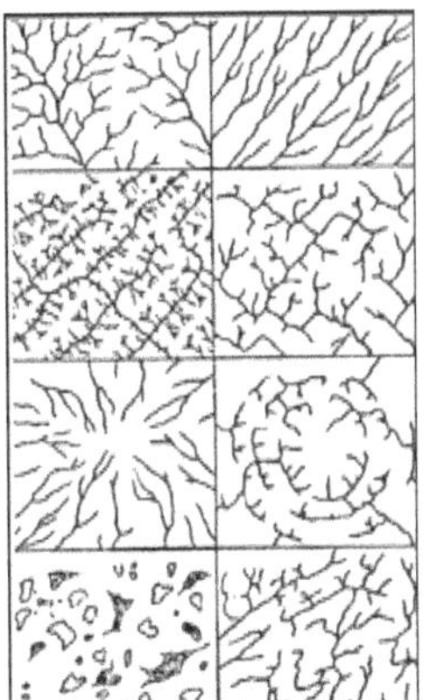

Fig. ***Major Types of Drainage Patterns***

The relationship of channel slope to sinuosity in an experimental river was elaborated by Schumm and Kahn. The data display a clear threshold phenomenon, in which steep low-sinuosity streams may change, somewhat abruptly, to somewhat less steep high-sinuosity streams. The former comprise the bedload-type streams that yield braided patterns, whereas the latter yield the familiar meandering patterns associated with streams that transport a high suspended load. The shift between these two stable pattern configurations is illustrated by several study areas, including the Yukon, Kosi, Pantanal, Japurá, and Ucayali.

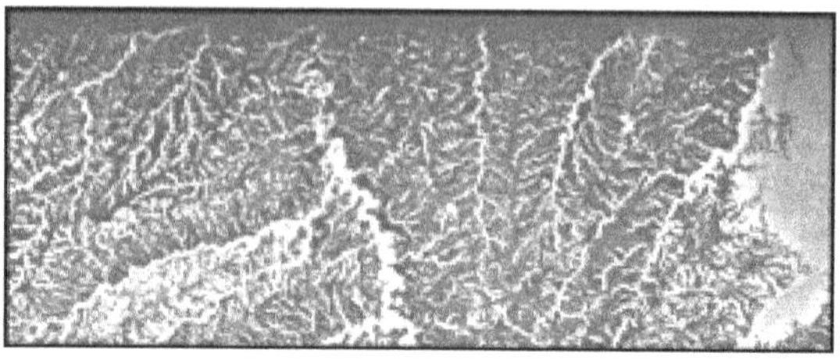

Fig. ***SIR-A Radar Image of Dendritic Drainage in Wast-central Columbia.***

Most of the image shows an area of dissected plains with a grassland cover that yields low radar return. The drainage pattern is strongly enhanced in radar return because the forested stream channels reflect the radar energy back to the receiver. On the basis of the foregoing experimental work, a variety of pattern classifications can be proposed. However, the immense complexity of natural fluvial systems appears to defy our present understanding.

Meandering Pattern

Meandering is the most common river pattern, and meandering rivers develop alternating bends with an irregular spacing along the valley trend. Such rivers tend to have relatively narrow, deep channels and stable banks. The system adjusts to varying discharge by vertical accretion on its floodplain and/or by lateral migration of its channel.

A vast complex of floodplain depositional features is associated with such rivers, as illustrated by the Mississippi River study area.

Braided Pattern

Braided rivers have channels divided into multiple thalwegs by alluvial islands. Braided rivers tend to have steeper gradients, more variable discharge, coarser sediment loads, and lower sinuosity than meandering streams. Their channels tend to be relatively wide and shallow. Braided patterns are "... developed depositionally within a channel in which the flow obstructions are sand and gravel deposited by the water moving around them". Midchannel bars are emplaced because of local flow incompetence.

The resulting braid channels formed by splitting the flow are more competent than the original channel for conveying the load downstream. Another way of describing braiding is that it is caused by channel widening that increases the boundary resistance of rivers with non-cohesive banks.

To maintain enough velocity for sediment transport in a wide, shallow cross section, the channel must divide and form relatively narrow and deep secondary channels through incision. Excellent examples of braiding occur in gravel-transporting rivers, such as Yukon, Colville, and upper Kosi. Braiding can also occur in sand-transporting rivers, like the Brahmaputra. The latter experience more frequent and more complex modification of original bar forms.

Anastomosed Pattern

Many multichannel rivers have relatively low gradients, deep and narrow channels, and stable banks. Such river systems have been termed "anastomosed". The terminology is a bit confused because "anastomosis" is a general designation for interconnected channelways whether in alluvial or in bedrock rivers.

Thus, Garner, following Bretz defined an anastomosing channel system as "... an erosionally developed network of channels in which the insular flow obstructions represent relict topographic highs and often consist of bedrock." Anastomosis is extensively developed in the Channeled Scabland.

Therefore, anastomosing patterns can be considered to be composed of multiple interconnecting channels separated by relatively stable areas of floodplain or bedrock. In contrast, braided patterns are single-channel, multiple-thalweg systems with bars of sediment or vegetated islands around which flow is diverted in the channel.

Excellent examples of anastomosed streams occur in the plainslands of east-central Australia. The Burke and Hamilton Rivers and the Cooper Creek study areas illustrate these arid-region varieties. Anastomosis also characterizes very large tropical rivers, such as those in the Amazon Basin. The Solimões and Japurá study areas illustrate such rivers.

Distributary Pattern

Distributary patterns occur where fluvial systems are spreading water and sediment across depositional basins. Two varieties are fans and deltas.

Fans develop in piedmont areas under the influence of both tectonic and climatic controls. Arid-region alluvial fans are constructed by infrequent depositional events that include both debris flows and water flows. Typical arid-region fans occur in the Tucson and Tian Shan study areas. Cold-climate alluvial fans occur in areas of glacial outwash and in periglacial regions. An excellent example is the Sheenjek Fan in the Yukon River study area. Humid-region alluvial fans are constructed by seasonal or perennial fluvial flows. The Kosi Fan of Nepal and India is an example from an area of active mountain building. The Pantanal study area illustrates some large fans in the savanna tropics of Brazil.

Deltas are the subject of another stage in this volume since most deltas involve the interaction of ponded water systems with sediment delivered to a river mouth. However, some basins of deposition in arid regions lack ponded water. Rivers entering these basins may produce typical deltaic morphologies, as in the case of the Niger River study area in Mali, West Africa.

Transitional Patterns

Five study areas from the Amazon Basin illustrate the complexity of tropical river systems. Many of these complexities arise because the fluvial system is not merely an entity that is totally adjusted to the vagaries of modern conditions. Rivers possess a heritage in which they inherit elements of ancient conditions.

Thus, old buried structures, relict alluvium, and progressive development contribute detail to the modern fluvial landscape. The understanding of modern rivers requires an understanding of their past history.

TECTONIC LANDFORMS

All continents are part of crustal plates and have two common components; cratons and folded linear mountain belts. Cratons are expansive, stable regions of low relief typically in the central part of the continent and made up of old igneous and metamorphic rocks buried under a relatively thin mantle of sedimentary rocks. The key part of the concept of a craton is stable. Although the old rocks forming the core may have been extensively metamorphosed in the geologic past, the craton has not undergone appreciable metamorphism or rapid uplift for several hundreds of million years. If other processes have not modified the sedimentary veneer of the cration, soils on the craton can be very old because of low relief and low erosion rates. The northern Great Plains of the U.S. and central Canada form the North American craton. Folded linear mountain belts are common on the margins of continents. Their occurrence is related to current or past collosions of plates. Rocks are typically extensively metamorphosed with intrusions of igneous rocks (Stone Mountain). Because relief is usually high, high rates of erosion prevent development of very mature soils. Both the Appalachian and Rocky Mountains are examples of folded linear mountain belts. The Appalachian Mountain chain formed about 300-350,000,000 years ago as the result of the collision of the North American and African plates.

The Piedmont of the eastern U.S. is part of this mountain chain, but erosion has worn down the mountains to the rolling landscape we see today. The Blue Ridge and other higher mountain chains further inland were folded and uplifted slightly later that the Piedmont and have not eroded as much. The Rocky Mountains formed from the collision of the North American and Pacific plates during the Tertiary epoch (65-1.8 million years before present (ybp)). Because these mountains are considerably younger than the Appalachians, they are higher and steeper. Other examples include the Himalayan Mountains, the Alps, and the Andes. Volcanism is also common during this type of mountain building. Volcanic deposits are common in the western U.S. and have been identified in the Appalachians and Piedmont.

Glacial Landforms

Climate change is the rule rather than the exception during the earth's history. Continental glaciers that covered much of the high latitude regions of the earth during ice ages have sculpted most of the landforms in these areas. During the Pleistocene epoch (1,800,000 to 12,000 ypb), there were multiple glacial advances and these glaciers covered much of the earth's higher latitudes. In the U.S., glaciers extended as far south as St. Louis, MO during periods of maximum glaciation. Glaciers form a wide variety of landforms, and glacial deposits are widely variable. Till, the most widespread type of glacial deposit, was the material pushed, churned, and modified under the glacial ice.

An analogy is a huge bulldozer that pushes all the material across the landscape and the resulting deposit is a mixture of whatever was in the path. Till is typically poorly sorted with particle sizes that range from clay to boulders. Composition of the particles depends on what was present in the path of the glacier. Composition of till in the midwestern U.S. is different than that in the northeast because the rock types encountered by the glacier were different. One type of deposit not formed by the glacier but that is associated with glaciers is loess; silty eolian materials. Even during glaciation, there was winter and summer. During summer, glaciers melted and large quantities of meltwaters flowed away from the glacier in steams. Along with the water were large quantities of sediment. The result was large stream discharges with large quantities of sediment in major rivers such as the present day Ohio, Missouri, and Mississippi rivers. The river channels were very broad to accommodate the flow and large amounts of sediment were deposited in the channel and on the broad floodplain.

During winter, the glaciers melted less, and the flow and associated sediment load in the rivers was substantially reduced. The reduced stream flow left large areas of freshly deposited sediment uncovered by water and with no vegetation. This sediment was entrained by wind and deposited in the adjacent uplands where vegetation was present to trap the eolian sediments. Sands were deposited near the channel as eolian dunes. The finer silt and clay were transported further from the channel before being deposited as what we now refer to as loess. Loess deposits are extensive in regions that carried glacial meltwaters including the central U.S., northern Europe, and northern Asia. Because it is an eolian deposit, loess is a blanket that covers all existing landforms and rock types. Thickness is uniform on all surfaces locally, but thickness of the loess deposit decreases with distance away from the river source. Near the river valley, the loess may be many meters thick and gradually thins with distance from the source. Because the Midwest U.S. had numerous stream sources, most of this part of the U.S. is has a loess cap that is <1 to several meters thick and blankets whatever rock or sediment that was present (till, sedimentary rocks and deposits). Along the Mississippi River, whose ancestor drained most of the glaciated region of the U.S., thick loess deposits occur as far south as central Mississippi. The loess is several meters thick near the river and thins away from the river to the point that it cannot be identified more than 80-100 km from the river. Soils associated with glacial landforms have a known age which cannot be older than the time of the glacial advance. The last glacial maximum in North America was about 12,000 ybp. Thus, soils in the central U.S. formed in either glacial deposits or loess are younger than 12,000 years. Soils formed in loess are silty and typically have <10 per cent sand.

"Marine" Landforms and Deposits

The term "marine" deposits is somewhat of a misnomer because most of the deposits we commonly refer to as marine as actually deposited at the sea-land interface. Thus, coastal deposits or continental shelf deposits would be a better term. Most continents have coastal regions, typically with relatively low relief, with unconsolidated sedimentary materials that were deposited in a variety of coastal environments including beaches, dunes, marshes, and deltas similar to those landforms that are currently on the coast. As stated earlier, climate change is the norm not the exception. Commonly associated with climate change is a change in sea level. Using Georgia as an example, during the Tertiary epoch, the ocean covered at least the southern half of the State, and some evidence suggests the sea level was as far north as Gainesville at one time during this period (Athens was under water). During the Pleistocene epoch, sea level was as much as 100 m lower than it is currently. In coastal plain regions (southern ½ of Georgia), particle size and composition of sediments (soil parent materials) vary with the environment in which they were deposited. For example, soils developed on ancient beach dunes are very sandy. Soils developed in ancient marshes are clayey and the list goes on. Since the sediments were transported to the coast by rivers, mineral composition of the sediments is strongly influenced by the mineralogy of the soils and sediments in the watersheds of the streams.

Limestone is another type of "marine" deposit that is composed of calcite ($CaCO_3$) or dolomite ($CaMg(CO_3)_2$). Most limestone was precipitated by marine organisms on the continental shelf some distance from the shoreline (think of coral reefs). Thus, limestone often contains a minimal amount of silicate minerals (quartz, feldspar, clay minerals, etc.). During the Cretaceous epoch (age of the dinosaurs; 146-65 million ybp), warm shallow seas (prime environment for carbonate formation and limestone deposition) covered extensive areas of North America (much of the central part of the continent) and thick limestone deposits occur in these areas. As limestone weathers and soil begins to form, the calcite and dolomite dissolve and are leached from the soil in ionic form. Thus, limestone derived soils are formed from the non-carbonate residues which are often clayey. Thus, soils over limestone in humid regions are often clayey and are also often red. In arid regions, limestone derived soils often have accumulations of calcium carbonate in the subsoil because of incomplete leaching of the carbonate. A question to ponder; if a soil derived from limestone that contained 5 per cent silicate residue is 2 m thick, how many feet of limestone were dissolved to form the soil?

UNDERSTANDING THE LANDSLIDERS

Landslides are simply defined as the mass movement of rock, debris or earth down a slope and have come to include a broad range of motions whereby falling, sliding and flowing under the influence of gravity dislodges earth material. They often take place in conjunction with earthquakes, floods and volcanoes. At times, prolonged rainfall causing heavy block the flow or river for quite some time. The formation of river blocks can cause havoc to the settlements downstream on it's bursting.

In the hilly terrain of India including the Himalayas, landslides have been a major and widely spread natural disaster the often strike life and property and occupy a position of major concern

One of the worst tragedies took place at Malpa Uttarkhand (UP) on 11th and 17th August 1998 when nearly 380 people were killed when massive landslides washed away the entire village. This included 60 pilgrims going to Lake Mansarovar in Tibet. Consequently various land reform measures have been initiated as mitigation measures.

The two regions most vulnerable to landslides are the Himalayas and the Western Ghats. The Himalayas mountain belt comprise of tectonically unstable younger geological formations subjected to severe seismic activity. The Western Ghats and nilgiris are geologically stable but have uplifted plateau margins influenced by neo- tectonic activity. Compared to Western Ghats region, the slides in the Himalayas region are huge and massive and in most cases the overburden along with the underlying litho logy is displaced during sliding particularly due to the seismic factor.

Landslides are simply defined as the mass movement of rock, debris or earth down a slope and have come to include a broad range of motions whereby falling, sliding and flowing under the influence of gravity dislodges earth material. They often take place in conjunction with earthquakes, floods and volcanoes. At times, prolonged rainfall causing heavy block the flow or river for quite some time. The formation of river blocks can cause havoc to the settlements downstream on it's bursting. In the hilly terrain of India including the Himalayas, landslides have been a major and widely spread natural disaster the often strike life and property and occupy a position of major concern

One of the worst tragedies took place at Malpa Uttarkhand (UP) on 11th and 17th August 1998 when nearly 380 people were killed when massive landslides washed away the entire village. This included 60 pilgrims going to Lake Mansarovar in Tibet. Consequently various land reform measures have been initiated as mitigation measures.

The two regions most vulnerable to landslides are the Himalayas and the Western Ghats. The Himalayas mountain belt comprise of tectonically unstable younger geological formations subjected to severe seismic activity. The Western Ghats and nilgiris are geologically stable but have uplifted plateau margins influenced by neo- tectonic activity. Compared to Western Ghats region, the slides in the Himalayas region are huge and massive and in most cases the overburden along with the underlying litho logy is displaced during sliding particularly due to the seismic factor.

PLATE TECTONIC FORCES

Plate tectonic is the idea that plates carry the continents and are great slabs of solid material that make up the ocean floor. Plate tectonics comes from the Greek word,"tektonikos" meaning "builder." It has been determined that there are about 20 rigid plates that are in slow, continuous motion. Some continents move at a rate of 1/2 to 4 inches per year which is directed by heat driven convection cells in the molten rock deep below the crust. As they move, they carry the continents and ocean floor.

In the late 1800's, Alfred Wegener, a German physical geographer, used spatial analysis to propose the continental drift hypothesis. Wegener studied the outlines of the continents and suggested that the existing land masses had been united at one point in the earths early history. He called his theory"die Verschiebung der Continent" meaning "continental displacement." His idea stated that these stable, immovable continents were mobile with the help of the tectonic plates. With further research his theory was accepted, but not until 60 to 70 years later. The Earth is made up of three layers: the crust, the mantle and the core.

The crust is a thin (15 mile) layer covering the outside part of the earth. The second layer is the mantle which is 1,800 miles thick. The crust and the upper mantle make up what is called the lithosphere. This lithosphere is 60-90 miles below the continents and 40-50 miles below the oceans. The plates in the plate tectonic theory are the lithosphere. The continental crust is less dense or lighter than the oceanic crust and "floats" above it. The base of the lithosphere is called the asthenosphere. Here is where the lithosphere is unattached from the mantle and moves around, mostly by gravity and thermal differences in the mantle.

The core, the third layer, is 1,000 miles thick. The core and the mantle are made of hot molten rocks, but the core is much hotter than the mantle. Below the 15 mile crust there is an increased amount of heat. It is believed that this heat is 'left over' from the formation of the earth and decaying radioactive material is fueling the fire. In fact, it is a possibility we could use the energy from this heat to fuel our lives if we were to run out of oil. How does this heat cause the plates to move? The earth's crust is cold, the mantle is hot and the core is even hotter, thus providing us with the explanation. In order to equalize these temperatures, convection cells are formed. Two types of rotation are produced by these convection cells.

Propelled by these heat convection cells, these plates move very slowly; one to four inches per year. (A couple of inches per year isn't much, since the earth's history is measured in millions of years). 225 million years ago, it is postulated that a giant continent called Pangaea ('all-earth') existed. This giant continent remained until about 135 million years ago, when it began to break up during the Mesozoic time. The break up consisted of India detaching itself from Africa and Antarctica and headed into the Indian Ocean. A giant mountain range is formed where the Australian-Indian Plate is pushing into the heart of Asia.

Additional evidence supporting the continental drift theory is supplied by the amphibian and reptile fossils that are spread out among the widely separated continents. And the evidence of polar wandering and evidence of magnetic field reversals locked into oceanic basalt samples. There are only three ways plates can interact while moving. This, in turn, causes there to be only three types of boundaries which are produced by different stress fields. The first boundary is called divergent. This is a tension or a stress that pulls the plates apart.

Divergent boundaries cause mid oceanic ridges. Some other common characteristics are high heat flow, mild volcanic activity, and shallow earthquakes. The second boundary is called convergent and this is a compression, or a stress that can shorten or compresses the plates. Convergent boundaries cause mountain ranges to develop. The Himalayas for example, were formed when the plate carrying India collided with the plate carrying Eurasia. This continental collision is still active and moving at a rate of 3 to 4 inches per year as the India plate is pushed under the Asian plate and the mountain continue to grow. Strong earthquake activity is very common with areas of convergent boundaries as well.

The third type of boundary is a transform boundary and this is when the plates slide past each other along faults causing mid-oceanic ridges and trenches. One plate may be forced down into the mantle under the other plate. When this occurs, a deep oceans trench forms.

The largest ocean form is the Mariana Trench in the Pacific Ocean southwest of Guam. Another example is the San Andreas Fault with is between the North American plate and the Pacific plate. Here, earthquakes are common but not volcanoes, and the earthquakes tend to outline the major plates.

The earth is constantly being shaped by the dynamics of the tectonicactivity and plates motion. Some of the present day's biggest mountains and ocean trenches are examples of the great power of these plates in motion. 'Hot Spots' is a generally accepted term used to explain the formation of islands in the middle of the Pacific such as the Hawaiian Islands, the Line Islands, and the Tuamotus. The Galapagos Islands are similar in fashion, though not as aligned, but are located off the coast of Ecuador. A Hot Spot is caused by the magma that rises or plumes from the core to the surface causing volcanoes by penetrating the mantle.

As the plate moves, it carries along the volcano that was formed. In it's place, a new one begins to form from the sea floor, while the hot spot stays in one place. Islands form in a "chain" as a result.

The Hawaiian Islands get younger from east to west in the chain. On the island of Hawaii, which is still over the hot spot, volcanoes remain very active. Other events related to this activity include earthquakes, volcanoes, and geothermic activity. Earthquakes are caused by abrupt easing of strains that have been built up along geologic faults and by volcanic action. The result of this is movement in the earth's surface. These vibrations can be felt in the locally affected areas and measured by scientific devices around the world. These plate cycles may also form volcanic activity.

For example, the Pacific Ocean is surrounded by a nearly continuous plate-collision zone, called the 'Ring of Fire', here Volcanoes are the results of the instability of this zone. Japan lies near the colliding edges of three plates, hence, earthquakes and volcanoes are a constant threat to the islands population.

TECTONICS, STRUCTURES AND FT THERMOCHRONOLOGY

Structural geology is the study in theory, in the laboratory and in the field of mineral and rock deformation at scales ranging from intracrystalline to continental. Tectonics is the study of the construction of the Earth - how large-scale processes of rock formation and deformation interact to create the planet. Research at Penn focuses on regional aspects of these sciences. Emphasis is placed on field study and laboratory analysis to develop regional deformational histories to aid in understanding the assembly of geologic regions. Investigations examine settings as varied as modem and ancient convergent plate boundaries (the Mariana forearc and the California Coast Ranges) and intracontinental deformation belts (basement uplifts in Iberia and in the Rocky Mountains). Within these regions, inquiries have included experimental deformation of serpentinite muds, studies of basalt geochemistry, fission-track analysis of basement and cover rocks, and seismic and gravity interpretation. Extensive collaboration with colleagues from other leading research institutions provide our students with access to a wide range of expertise, equipment and perspectives.

Among others, these institutions include the Lamont-Doherty Earth Observatory, the Smithsonian Institution, the University of Hawaii and the University of Lisbon.

FT Thermochronology: Fission-track thermochronology is one of the newest and most powerful tools geologists use to reconstruct tectonic and thermal histories of diverse geologic terranes. The method has been applied to many fields of geology, thus enhancing positive interactions between scientists in different disciplines of earth and planetary sciences. Research in this program involves the deciphering of tectonic and thermal histories of rift margins and basins, orogenic belts and sedimentary basins, and dating of meteorite impact events

Structural Geology and Tectonics:

1. Intracontinental Deformation: Spain, Portugal and the midcontinental U.S. Comparative investigation of the intracontinental consequences of collisional orogeny at continental margins. Such collisions produce deformation far into the continent when basement blocks are shifted along old faults, commonly those related to failed rifts.
2. Late Mesozoic Paleogeography of the U.S. Pacific Coast: California & Oregon. Current models do not explain the presence of a previously unmapped ophiolitic milange unit lying within the forearc of the Franciscan subduction complex but outside the complex itself. The modern Mariana subduction zone appears be a good actualistic analogue for many aspects of the fossil Late Mesozoic one along the U.S. Pacific coast. Investigations at Penn focus on testing this comparative model, and currently include several field projects along a spectacular section of the coast in southwestern Oregon.
3. Tectonics of the California Coast Ranges: Compressional stresses between the Pacific and North American plates that are not resolved by transcurrent motion along faults of the San Andreas system are being taken up by active thrust faulting and related folding, forming a variant of Valley-and-Ridge geometry containing exotic rock types such as ophiolites and forearc-basin subsea-fan rocks. Work in this area includes fission-track studies (in collaboration with Dr. Omar) and studies of fault geometry and timing.

Dr. Omar uses FT and thermal modeling to reconstruct tectonic and thermal histories of the following geologic terranes:

1. The Red Sea Rift: Timing, geometry, and extent of rift flank uplift and its relationship to subsidence history of the rift basin. Ultimate

objective is to obtain a greater understanding of the interaction of tectonics, geologic structure and erosion at the Red Sea rift margin.

2. Atlantic Margin Rift Basins: Newark and Taylorsville basins. Evaluation of the timing, spatial distribution, and migration of crustal-scale fluid-flow, a poorly understood phenomenon, the effects of which are only now becoming apparent.
3. Southeast Brazil: The goal of this project is to allow definitive limits to be placed on the timing of uplift and denudation of the southeast Brazilian topography and mechanisms for their formation.
4. Rocky Mountains: the main objective of this project is to determine the tectono-thermal history of individual basement blocks and intervening sedimentary basins in order to decipher the tectonic and deformation history of the western USA.

8

Hydrogeology, Geomorphology and Mapping System

Movement of water and solutes across a landscape depends on geomorphic and stratigraphic relationships including landscape distribution of soil horizons. Landscape configuration of soil and geomorphic surfaces affects surface flow pathways. In a similar manner, landscape configuration and distribution of soil horizons, especially those that are water restrictive, will influence paths for movement of water and solutes in the subsurface. Water runs downhill over the soil surface and as shallow subsurface lateral water movement. This is captured in Darcy's Law for saturated flow.

$$J = Q/A = K_s(dh/L)$$

Where:

J = flux, units of length/time

Q = volume/time, cm3/day for example

A = area

K_s = saturated hydraulic conductivity, units of length/time

dh = hydraulic head, units of length

L = length of flow path

Darcy's Law is most commonly applied when discussing vertical water movement through soils, but it is also applicable when considering rate of lateral movement of water across and through a landscape. Thus, in rolling landscapes with convex hillslope summits such as those in the Piedmont, soils on the lower part of the hillslope will be wetter than soils higher in the landscape. In the landscapes with the low relief and broad interfluves (summits), such as the Atlantic Coast Flatwoods, there is little gradient to move water laterally across or through the soil to streams draining the area.

As a result, soils in the central part of the interfluve have high seasonal water tables and are often poorly drained. Near streams, the gradient for lateral movement of water is greater and the soils are better drained. This relationship in these landscapes is often referred to as the "Dry Edge" or "Red Edge" effect (soils on edge of the interfluve are better drained and redder than those in the interior).

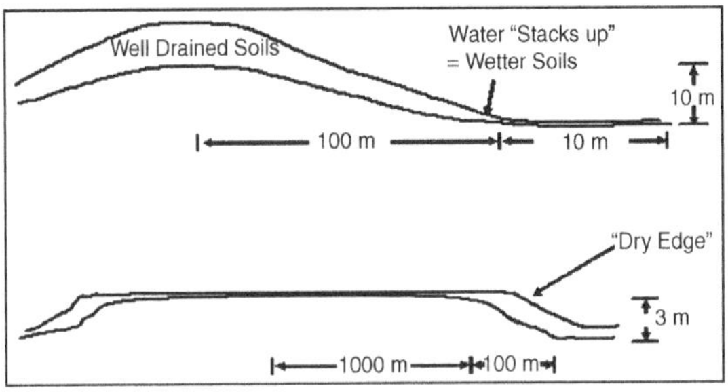

J = (20 cm/d) × (9.9 m/100 m) = 1.98 cm/d

J = (20 cm/d) × (0.1 m/10 m) = 0.2 cm/d

J = (20 cm/d) × (0.5 m/1000 m) = 0.1 cm/d

J = (20 cm/d) × (2.5 m/100 m) = 0.5 cm/d

As the width of the interfluve becomes less, the "edge" with sufficient gradient for lateral water movement to the stream comprises a greater proportion of the interfluve and a smaller proportion of the interfluve has poorly drained soils. The endproduct of interfluve narrowing is a rolling landscape in which summits are well drained and lower landscape positions are wetter.

In addition to differences in soil colour, seasonal water table depth also effects other soil properties. For example, greater depths to a seasonal water table results in greater E horizon thickness, more clayey B horizons, less clay in E horizons, decreased BE horizon thickness, and greater content of gibbsite in B horizons of soils in the Atlantic Coast Flatwoods.The primary reason for these changes in soil properties with depth to seasonal water table is increased water movement through the soil. If the water table is near the surface for much of the year, the soil is not leached because the water table limits vertical water movement.

HYDROGEOLOGY

Hydrogeology (*hydro-* meaning water, and *-geology* meaning the study of the Earth) is the area of geology that deals with the distribution and movement of groundwater in the soil and rocks of the Earth's crust (commonly in aquifers). The term geohydrology is often used interchangeably. Some make the minor distinction between a hydrologist or engineer applying themselves to geology (geohydrology), and a geologist applying themselves to hydrology(hydrogeology).

Hydrogeology is an interdisciplinary subject; it can be difficult to account fully for thechemical, physical, biological and even legal interactions between soil, water, natureand society. The study of the interaction between groundwater movement and geology can be quite complex. Groundwater does not always flow in the subsurface down-hill following the surface topography; groundwater follows pressure gradients(flow from high pressure to low) often following fractures and conduits in circuitous paths.

Taking into account the interplay of the different facets of a multi-component system often requires knowledge in several diverse fields at both the experimentaland theoretical levels. The following is a more traditional introduction to the methods and nomenclature of saturated subsurface hydrology, or simply the study of ground water content.

Hydrogeology in Relation to other Fields

Hydrogeology, as stated above, is a branch of the earth sciences dealing with the flow of water through aquifers and other shallow porous media (typically less than 450 m or 1,500 ft below the land surface.) The very shallow flow of water in the subsurface (the upper 3 m or 10 ft) is pertinent to the fields of soil science,agriculture and civil engineering, as well as to hydrogeology. The general flow of fluids (water, hydrocarbons, geothermal fluids, etc.) in deeper formations is also a concern of geologists, geophysicists and petroleum geologists. Groundwater is a slow-moving, viscous fluid (with a Reynolds number less than unity); many of the empirically derived laws of groundwater flow can be alternately derived in fluid mechanics from the special case of Stokes flow (viscosity and pressure terms, but no inertial term).

The mathematical relationships used to describe the flow of water through porous media are the diffusion and Laplace equations, which have applications in many diverse fields. Steady groundwater flow (Laplace equation) has been simulated using electrical, elastic and heat conduction analogies. Transient groundwater flow is analogous to the diffusion of heat in a solid, therefore some solutions to hydrological problems have been adapted from heat transferliterature.

Traditionally, the movement of groundwater has been studied separately from surface water, climatology, and even the chemical and microbiologicalaspects of hydrogeology (the processes are uncoupled). As the field of hydrogeology matures, the strong interactions between groundwater, surface water,water chemistry, soil moisture and even climate are becoming more clear.

Definitions and Material Properties

One of the main tasks a hydrogeologist typically performs is the prediction of future behavior of an aquifer system, based on analysis of past and present observations. Some hypothetical, but characteristic questions asked would be:

- Can the aquifer support another subdivision?
- Will the river dry up if the farmer doubles his irrigation?
- Did the chemicals from the dry cleaning facility travel through the aquifer to my well and make me sick?
- Will the plume of effluent leaving my neighbor's septic system flow to my drinking water well?

Most of these questions can be addressed through simulation of the hydrologic system (using numerical models or analytic equations). Accurate simulation of the aquifer system requires knowledge of the aquifer properties and boundary conditions. Therefore a common task of the hydrogeologist is determining aquifer properties using aquifer tests.

In order to further characterize aquifers and aquitards some primary and derived physical properties are introduced below. Aquifers are broadly classified as being either confined or unconfined (water table aquifers), and either saturated or unsaturated; the type of aquifer affects what properties control the flow of water in that medium (e.g., the release of water from storage for confined aquifers is related to the storativity, while it is related to the specific yield for unconfined aquifers).

Hydraulic Head

Differences in hydraulic head (h) cause water to move from one place to another; water flows from locations of high h to locations of low h. Hydraulic head is composed of pressure head (ψ) and elevation head (z). The head gradient is the change in hydraulic head per length of flowpath, and appears in Darcy's law as being proportional to the discharge.

Hydraulic head is a directly measurable property that can take on any value (because of the arbitrary datum involved in the z term); ψ can be measured with a pressure transducer (this value can be negative, e.g., suction, but is positive in saturated aquifers), and z can be measured relative to a surveyed datum (typically the top of the well casing). Commonly, in wells tapping unconfined aquifers the water level in a well is used as a proxy for hydraulic head, assuming there is no vertical gradient of pressure. Often only *changes* in hydraulic head through time are needed, so the constant elevation head term can be left out ($Dh = Dø$). A record of hydraulic head through time at a well is a hydrograph or, the changes in hydraulic head recorded during the pumping of a well in a test are called drawdown.

Porosity

Porosity (n) is a directly measurable aquifer property; it is a fraction between 0 and 1 indicating the amount of pore space between unconsolidated soilparticles or within a fractured rock. Typically, the majority of groundwater (and anything dissolved in it) moves through the porosity available to flow (sometimes called effective porosity). Permeability is an expression of the connectedness of the pores. For instance, an unfractured rock unit may have a high *porosity* (it has lots of *holes* between its constituent grains), but a low *permeability* (none of the pores are connected). An example of this phenomenon is pumice, which, when in its unfractured state, can make a poor aquifer.

Porosity does not directly affect the distribution of hydraulic head in an aquifer, but it has a very strong effect on the migration of dissolved contaminants, since it affects groundwater flow velocities through an inversely proportional relationship.

Water Content

Water content (q) is also a directly measurable property; it is the fraction of the total rock which is filled with liquid water. This is also a fraction between 0 and 1, but it must also be less than or equal to the total porosity.

The water content is very important in vadose zone hydrology, where the hydraulic conductivity is a strongly nonlinear function of water content; this complicates the solution of the unsaturated groundwater flow equation.

Hydraulic Conductivity

Hydraulic conductivity (K) and transmissivity (T) are indirect aquifer properties (they cannot be measured directly). T is the K integrated over the vertical thickness (b) of the aquifer (T=Kb when K is constant over the entire thickness). These properties are measures of an aquifer's ability to transmit water.Intrinsic permeability ($\hat{e}$) is a secondary medium property which does not depend on the viscosity and density of the fluid (K and T are specific to water); it is used more in the petroleum industry.

Specific Storage and Specific Yield

Specific storage (S_s) and its depth-integrated equivalent, storativity ($S=S_s b$), are indirect aquifer properties (they cannot be measured directly); they indicate the amount of groundwater released from storage due to a unit depressurization of a confined aquifer. They are fractions between 0 and 1.

Specific yield (S_y) is also a ratio between 0 and 1 (S_y d" porosity) and indicates the amount of water released due to drainage from lowering the water table in an unconfined aquifer. The value for specific yield is less than the value for porosity because some water will remain in the medium even after drainage due to intermolecular forces. Often the porosity or effective porosity is used as an upper bound to the specific yield. Typically S_y is orders of magnitude larger than S_s.

Contaminant Transport Properties

Often we are interested in how the moving groundwater will transport dissolved contaminants around (the sub-field of contaminant hydrogeology). The contaminants can be man-made (e.g., petroleum products, nitrate or Chromium) or naturally occurring (e.g., arsenic, salinity). Besides needing to understand where the groundwater is flowing, based on the other hydrologic properties, there are additional aquifer properties which affect how dissolved contaminants move with groundwater.

Hydrodynamic Dispersion

Hydrodynamic dispersivity (α_L, α_T) is an empirical factor which quantifies how much contaminants stray away from the path of the groundwater which is carrying it. Some of the contaminants will be "behind" or "ahead" the mean groundwater, giving rise to a longitudinal dispersivity (α_L), and some will be "to the sides of" the pure advective groundwater flow, leading to a transverse dispersivity (α_T). Dispersion in groundwater arises because each water "particle", passing beyond a soil particle, must choose where to go, whether left or right or up or down, so that the water "particles" (and their solute) are gradually spread in all directions around the mean path. This is the "microscopic" mechanism, on the scale of soil particles. More important, on long distances, can be the macroscopic inhomogeneities of the aquifer, which can have regions of larger or smaller permeability, so that some water can find a preferential path in one direction, some other in a different direction, so that the contaminant can be spread in a completely irregular way, like in a (three-dimensional) delta of a river.

Dispersivity is actually a factor which represents our *lack of information* about the system we are simulating. There are many small details about the aquifer which are being averaged when using a macroscopic approach (e.g., tiny beds of gravel and clay in sand aquifers), they manifest themselves as an*apparent* dispersivity. Because of this, α is often claimed to be dependent on the length scale of the problem — the dispersivity found for transport through 1 m^3 of aquifer is different than that for transport through 1 cm^3 of the same aquifer material.

Molecular Diffusion

Diffusion is a fundamental physical phenomenon, which Einstein characterized as Brownian motion, that describes the random thermal movement of molecules and small particles in gases and liquids. It is an important phenomenon for small distances (it is essential for the achievement of thermodynamic equilibria), but, as the time necessary to cover a distance by diffusion is proportional to the square of the distance itself, it is ineffective for spreading a solute over macroscopic distances. The diffusion coefficient, D, is typically quite small, and its effect can often be considered negligible (unless groundwater flow velocities are extremely low, as they are in clay aquitards).

It is important not to confuse diffusion with dispersion, as the former is a physical phenomenon and the latter is an empirical factor which is cast into a similar form as diffusion, because we already know how to solve that problem.

Retardation by Adsorption

The retardation factor is another very important feature that make the motion of the contaminant to deviate from the average groundwater motion. It is analogous to the retardation factor of chromatography. Unlike diffusion and dispersion, which simply spread the contaminant, the retardation factor changes its *global average velocity*, so that it can be much slower than that of water. This is due to a chemico-physical effect: the adsorption to the soil, which holds the contaminant back and does not allow it to progress until the quantity corresponding to the chemical adsorption equilibrium has been adsorbed. This effect is particularly important for less soluble contaminants, which thus can move even hundreds or thousands times slower than water. The effect of this phenomenon is that only more soluble species can cover long distances. The retardation factor depends on the chemical nature of both the contaminant and the aquifer.

GEOMORPHOLOGY

Geomorphology the science that studies the evolution of topographic features by physical and chemical processes operating at or near the earth's surface. In geomorphology, the landscape is viewed as an assemblage of landforms which are individually transformed by landscape evolution. Because soils are an integral part of the landforms and landscape, processes occurring on the landscape have implications for soil processes. Conversely, soil processes can be considered to be a part of landscape evolution.

Landscape: the portion of the land surface that the eye can comprehend in a single view.

Landforms: distinctive geometric configurations of the earth's surface; features of the earth that together comprise the land surface

Geomorphic surface: a part of the surface of the land that has definite geographic boundaries and is formed by one or more agents during a given time span. It should be considered as a surface, *i.e.* similar to a plane, no thickness (z) - only x and y dimensions. Because it is formed during a specific time it is datable, either by absolute or relative means.

Erosion surface: a land surface shaped by the action of ice, wind, and water; a land surface shaped by the action of erosion

Constructional (depositional) surface: a land surface owing its character to the process of upbuilding, such as accumulation by deposition, for example fluvial, colluvial, or eolian deposits).

Geomorphic principles

Practically all work dealing with the distribution of soils on the earth's surface employs geomorphic concepts, and it is especially important in two areas;

1. Age, properties, and development rate of soils and
2. Hydrologic patterns on landscapes.

Soil Age

Soil development does not commence until erosion or deposition rate has reached a steady state that is less than the rate of soil formation. Thus, the age of the soil may or not be similar to the age of the deposits underlying the soil. For depositional surfaces, soil age is similar to the age of deposit (normally, slightly younger but related to the depositional event).

Thus, radiocarbon or other dating methods of materials in the deposit are useful for determining soil age. This is not true for erosional surfaces. There may have been multiple erosion episodes since the material was deposited or exposed to sub-aerial processes. Thus, often the best that can be done is a relative age of the surface compared to other geomorphic surfaces in the area.

TOPOGRAPHY AND GEOLOGIC MAPPING

A dipping bed can be projected along a contour only when the trend of the hillslope is parallel to the strike. If the strike does not parallel the contours, the trace of the bed along the hill will lie either higher or lower than the first outcrop, depending on whether the bed is followed in the up-dip or down-dip direction. This simple geometrical fact is the basis for most determinations of attitude of beds with low dips, for it is difficult to read a clinometer more closely than 1°. It is also the basic clue to reading a geologic map in order to glean from it the succession of the beds and their structure.

In the Paris Basin most of the streams drain to the Seine. East of Paris they flow westward, cutting approximately at right angles through the arcuate ridges. Applying the relation between the dip of a bed and its trace on the land surface, we find that, if we select some stratum such as one of the thin layers of greensand in the Lower Chalk, or one of the thin limestones that lies above the Plastic Clay, and trace the outcrops of this bed westward along the banks of a stream, the bed decreases gradually in elevation when followed to the west until it reaches the level of the stream; thereupon it crosses the stream bed, reverses its direction and climbs higher and higher when followed eastward on the opposite bank of the stream.

In other words, the outcrop pattern is a "V" with the point of the "V" directed downstream. This can only be explained by the fact that the bed is inclined westward, and that its dip angle is greater than the fall (gradient) of the stream.

The fact that the Upper Chalk crops out further west in the valley of the Marne and Aisne rivers than it does on the intervening divides proves that the contact of the chalk with the overlying Plastic Clay also dips to the west. By using such facts the east-west section forming the lower edge (bottom) was drawn, although each formation shown on this section is projected to much greater depths within the earth than we can actually observe in the field. The basic principle of original continuity is the same as for horizontal beds, but in projecting the bed to a geologic section, account must be taken of the dip of the beds.

Relation of Topography to Underlying Rocks

There is generally a relationship, faithful in areas of little soil but more generalized in areas of deeper soil, between the surface form of a country and the underlying bedrock. In the Paris Basin the curving ridges mark outcrops of resistant strata such as sandstones. Porous chalk layers and other easily eroded rocks are found in the lowlands. Weathering and erosion have etched the surface into relief, forming lowlands on the less resistant beds and leaving the more resistant ones standing as hills.

The curving pattern of these hills and lowlands shows that the rocks of the Paris Basin have been warped into a shallow saucerlike basin. Indeed, the whole succession of strata resembles a pile of saucers of diminishing size, with Paris near the center of the smallest and uppermost saucer.

The correlation between topography and structure in the Paris Basin is by no means perfect—much closer adjustment is noted in deserts where the soil cover is thin or absent. However, the topography usually indicates the general trend of the underlying rocks. Generally, sandstone is more resistant to erosion than shale and hence forms a ridge. It is also more permeable to rain so that, if it rests on shale, there is likely to be a line of springs or a strip of flourishing plants along its lower contact.

Thus, the minor topographic features of an area nearly always furnish clues to the characters and trend of the underlying bedrock. If we find conglomerate on top of a low ridge at one point, it suggests that conglomerate may underlie the ridge throughout its length.

This is easily checked by seeking other outcrops along its summit. A road may expose chalk in a lowland. Perhaps a well has been dug in the same lowland a thousand feet away. Does it show chalk on its walls? By piecing together such information, the geologist can generally establish the stratigraphic succession, discover the variations in thickness of beds, and ultimately plot their positions on a map.

Even in country with a thick soil it is still possible to apply the principles of geologic mapping and correlation by using data gained from excavations, well borings, or geophysical methods.

Limitations of Scale

As in all other maps, geologic maps require the rigorous selection of data; they emphasize some features at the expense of others. The smaller the scale of the map, the fewer the details that can be shown. On which shows only a few square miles of the Paris Basin near Epernay, details of thin layers above the chalk are shown that could not possibly be indicated on the small-scale map of the entire Paris Basin prepared by Cuvier and Brongniart.

Thus the geologist must select his *geologic map units*, or *formations*, to fit the scale of the map. The larger the scale and the better the exposures, the larger the number of formations that can be shown within a given area, although the number per square inch of map remains about the same.

Formations

The basic unit of the geologic map is the *formation. There are two criteria for selecting a formation: first, its contacts (i.e., the top and the bottom of a sedimentary formation) must be recognizable and capable of being traced in the field, and second, it must be large enough to be shown on the map.*

In their studies of the Paris Basin, Cuvier and Brongniart noticed faint stratification surfaces within the Lower Chalk. The chalk above and below these stratification surfaces, however, was so nearly identical in appearance and fossil content that such subdivisions of the strata were not regarded by the two geologists as either significant or capable of being successfully traced and mapped in the field.

On the other hand, the contact between the Upper Chalk and the overlying Plastic Clay was mapped as a formation boundary because of the marked contrast in the rocks. A still more cogent reason for selecting this contact to map was the evidence that it represented a considerable interval of geologic time—enough for the underlying chalk to have become firmly coherent, so that it could be incorporated as pebbles and fragments in the lowermost bed of clay. The stratification within the clay was more prominent than the chalk, but again the similarity of each bed of clay to its neighbors made it difficult, if not impossible, to trace any particular stratification plane far.

The beds above the Plastic Clay posed a somewhat different problem to Cuvier and Brongniart. Here, there were many different kinds of rock in relatively thin layers—limestone, shales, sandstones, beds of gypsum, and clay. At some outcrops an individual bed, perhaps a clay only a foot thick, could be seen to thin and "lens out" within a few tens of feet.

Or the overlying sandstone, perhaps 15 feet thick in this outcrop, could be seen to thicken when followed across country in successive outcrops, and then to thin again and perhaps ultimately disappear. Only on a very large scale could each thin layer be shown on a map, and it would take a prodigious amount of time to trace the contacts. Such a series of thin, variable beds, often including very diverse rock types, were grouped together by Cuvier and Brongniart as a single formation.

Although the thin beds of such variable formations may be individually lenticular, and indistinguishable either from other beds of the same formation or adjacent ones, nevertheless the group as a whole is recognizably different from formations above and below it.

Depending on the scale of the map, the abundance of exposures, the character of the beds, and, not least, the discrimination of the geologist, very different map units (formations) might be selected in a certain section.

Any differences are adequate to justify calling a particular bed, or any closely related group of beds, a formation, provided the differences allow the formation to be recognized in scattered outcrops, and provided the top and bottom of the formation can be traced in the field. The purpose of the map and its scale largely determine what units to select as formations.

Geologic formations in the United States are nearly always given a place name from some geographic locality near which the formation was first identified. This is followed by the name of the *dominant* rock variety composing it, or, if it is composed of a great variety of rocks, by the word "formation." Thus: the *Austin chalk*, the *Columbia River basalt*, the *Chattanooga shale*, the *Denver formation*.

In Europe the practice is less formal; many formations are named from some characteristic fossil (the *Lingula flags*, a thin-bedded sandstone containing abundant fossils of the genus *Lingula*), some economic character (*Millstone grit*), or even a folk name (*Norwich Crag*).

GEOMORPHOLOGICAL MAPPING

Most of the interest in geomorphological mapping has centered on the development of various mapping systems for use in environmental management. The most detailed systems have been developed in Europe, where different countries utilize different procedures. Despite attempts at international standardization, the major problem remains the correlation of various mapping schemes.

Global remote sensing has great promise for generating geomorphic maps at large spatial scales. The use of the SIR-A system to define terrain categories for geomorphic mapping in Indonesia. The terrain types are recognized through interpretation of radar interaction with the ground surface, especially the surface roughness, vegetation, and topography. The radar information is made even more remarkable because the mapped areas are characterized by dense tropical forest cover and persistent clouds.

Future Directions

In 1982 in Vienna, the United States proposed at the United Nations Conference on Peaceful Uses of Outer Space that an international cooperative research programme be organized to understand the Earth as a system. This programme was initially named "Global Habitability" and was formulated to involve the central role by NASA in the observation of system parameters and changes.

In a sense, the programme constitutes a "mission to planet Earth" in which spaceborne remote sensing is applied to studies of dynamic processes in the atmosphere, biosphere, geosphere, and hydrosphere. The Global Habitability concept will probably be integrated into the broader efforts of the proposed International Geosphere-Biosphere Programme to be coordinated by the International Council of Scientific Unions.

In his keynote address to the World Conference on Earthquake Engineering, the President of the National Academy of Sciences proposed an "International Decade of Hazard Reduction". A major component of such an initiative must be the global analysis of hazardous geomorphic processes. What do such initiatives mean for geomorphology?

Global Tectonic and Climatic Systems

Tectonic geomorphology involves the interactions among landforms, landscapes, and tectonics. Tectonics, the branch of geology dealing with regional structures and deformation features, occupies a central role in the Earth sciences. The discipline has achieved great importance through the unifying role of the plate tectonic model in explaining the large-scale surface features of our planet.

Major advances in tectonic geomorphology have been made in the last decade, mainly because of an increased ability to evaluate the time factor in landscape development. Thus, through the use of geochemical means of dating and computerized models of landform change, it is now possible to evaluate differential rates of uplift.

It is also possible to determine the magnitude and frequency of displacements along faults. Perhaps the most important new development for climatic and climatogenetic geomorphology is the use of analytical models to characterize the global interactions of the land surface, atmosphere, and oceans. Most interesting are the general circulation models that simulate global atmospheric processes. For climatic geomorphology, these models have shown profound feedback relationships operating between soil moisture and precipitation, carbon dioxide and climate and anthropogenic changes in the albedo and climate.

An example of an especially useful climate-genetic reference point is provided by the CLIMAP reconstruction of the 18000 years B.P. global climate from a compilation of the Earth´s ocean surface temperatures. A preliminary extension of the analysis to continental areas revealed numerous problems of local variability.

Such comparisons between general atmospheric conditions, as modeled by GCMs, and paleogeomorphic reconstructions hold great promise for understanding large-scale climate/landscape interactions. The goal here is to generate a self-enhancing spiral of understanding, with models pointing to key geomorphic questions and geomorphic data refining the models.

Ancient Landscapes

At the end of the l9th century, geomorphology achieved an important theoretical synthesis through the work of William Morris Davis. Davis conceived a marvelous deductive scheme of landscape development by the action of exogenetic processes acting on the basic materials and structures to produce a progressive evolution of landscape stages through time.

Unfortunately, this theoretical framework was somewhat abused by those who employed it solely for landscape description and classification. By the middle of the 20th century, evolutionary geomorphology fell from favour among Earth scientists, who focused their primary efforts on the study of various geomorphic processes.

An unfortunate by-product of the controversy over Davisian geomorphology was the general abandonment, especially in Britain and the United States, of studies that concerned ancient landscapes and landforms. The development of radiometric dating has rekindled interest in this topic by identifying the antiquity of landscapes.

For example, Young has shown that upland surfaces in southeastern Australia originated as early as Mesozoic and Early Tertiary, with the landscape assuming its approximate present-day form by the Miocene. Twidale et al. also identified Mesozoic landscapes in South Australia. Ollier in an analysis of ancient landscapes in Australia, concluded that conventional approaches to geomorphic change suffer from inadequate appreciation of broad scales of time and space. He proposed an evolutionary approach to geomorphic features, such as that applied by geologists to tectonic features. Low-relief plains cutting across varied rocks and structures are common features on the Earth. Such surfaces have long been of interest to geomorphologists, and many scientific controversies have arisen over their explanation.

The genetic implications are contained in the many names for the surfaces; peneplains, pediplains, panplains, etchplains, exhumed plains, and paleoplains. To avoid the problems inherent in these controversies, it is perhaps best to simply call these features planation surfaces.

Table. Widespread Planation Surfaces

Name	Age	Comments
Gondwana	Jurassic	Related to Pangea and its breakup.
Post-Gondwana or Kretacic	Early to Mid-Cretaceous	Related to Pangea and its breakup.
African or Moorland	Late cretaceous to Early Cenozoic	Extensive surface created by stripping weathered material from older surfaces.
Post-African or Rolling	Miocene	Undulating surface developed above younger valleys
Widespread	Pliocene	Global surface common near coastal areas.
Youngest	Quaternary	Latest valley formation.

A spectacular planation surface that bevels sandstone cuestas in the fold belt of the Amadeus Basin in central Australia. An analysis of the regional geomorphology of this area is presented in Plate I-4.

The question of regional planation surfaces obviously awaits a modern global analysis.

Perhaps the classic syntheses of King can be re-evaluated by the use of the new techniques described in this volume. It would be refreshing to accomplish such a study with automated data collection procedures, free of the raging controversy that so hampered the highly personal interpretive studies of the past.

The identification and dating of various planation surfaces have become important components of tectonic geomorphic analysis. Deformation of such surfaces by faulting, folding, or broad warping can be calibrated by the displacement of a planation surface from its original attitude.

METHODS AND PRINCIPLES OF GEOLOGIC MAPPING

This brief account of the early geologic maps gives us glimpses and hints of the methods and principles used in geologic mapping. We should now summarize these principles, consider their validity, and show how they are applied in actual field work.

On William Smith's geologic map of England, lines representing the contacts between different rock formations are drawn for distances equivalent to hundreds of miles on the ground. Yet, in tracing an individual contact for 100 miles, Smith probably found, on the average, less than 50 exposures where the actual contact could be seen on a clean rock face. How, then, can his map record the real distribution of the rocks? Can it be anything but a guess?

The eyes of a geologist are no more capable of seeing the bedrock through a cover of soil than those of any other observer. How, then, can the geologist make inferences about the position of the bedrock underground that will withstand objective tests, such as those provided by digging wells and sinking mine shafts?

The succession must be pieced together from scattered outcrops. Although in an area of a few square miles a geologist may find only one or two outcrops in which he can see the actual contact between two formations, he will doubtless find 100 or more outcrops composed entirely of rock belonging to one or the other formation.

He thus has some clue as to the position of the contact with respect to each outcrop. The problem is roughly analogous to drawing a contour line to conform to the elevations determined at 100 or more control points. To use scattered rock outcrops, however, requires that the individual formations be correctly identified (correlated) from outcrop to outcrop, and this is not always easy.

Four Fundamental Postulates of Geologic Mapping

There are four fundamental postulates that underlie the making of a geologic map. The third is merely a commonsense deduction from these two, and, like them, was first stated by Steno: A water-laid stratum, at the time it was formed, must continue laterally in all directions until it thins out as a result of non-deposition, or until it abuts against the edge of the original basin of deposition. This is the Law of Original Continuity. An important corollary of this law, not fully appreciated by Steno in 1669, but well known to the geologists of France and England at the beginning of the Nineteenth Century, may be stated: A stratum which ends abruptly at some point other than the edge of the basin in which it was deposited, must have had its original continuation removed by erosion, or else displaced by a fracture in the earth's crust. These four postulates—

- Superposition (the higher bed is the younger);
- Original horizontality (stratification planes are formed roughly parallel to the earth's surface);
- Original continuity;
- Truncation by erosion or dislocation—are the basis for many of our interpretations of the relative relations of strata.

They are not absolute rules that can be rigidly applied. For instance, some beds, once horizontal, have been highly tilted and even overturned by movements of the earth's crust , so that a stratum formerly beneath another may now lie inverted on it; other strata, as at the front of a steep delta, may have been deposited on appreciable slopes; the edges of many landslides end abruptly instead of thinning to a feather edge. Such exceptions, however, are not common and can generally be recognized easily by the geologist. How do we correlate between scattered outcrops when we are in the field making a geologic map?

Correlation of Rock Outcrops

In a ravine on a grassy hillside we may see a bed of clay with well-marked horizontal stratification; we assume it continues horizontally into the hill at the same elevation, for how else can its horizontal stratification be extended?

If we go 200 yards along the same contour, without seeing an outcrop, and then find a clay bed at the same elevation in another ravine, we may suspect it is the same bed. If both are gray, the probability is heightened. If both show lines of nodular lumps (concretions) along the stratification planes, and if the size and spacing of these lumps is about the same, we are still more confident of our correlation.

If both rest on red limestone, both are overlain by fine-grained brown sandstone, and both contain the same kinds of fossils, we become practically certain of the correlation. We are now justified in assuming that the bed is continuous beneath the soil between the two exposures.

If we go on a little farther and come to a deep ravine in the hill which shows a very large outcrop containing not a few feet of strata but several hundred, and if in this section there is only one clay bed with features identical to those we saw in the two small outcrops, we have still further evidence that our mapping is correct. We can now map the clay, for the contacts of the bed with those above and below will parallel the contours, as they are horizontal planes along the top and bottom of the clay bed.

We have, furthermore, gained information on the thickness and relations of the red limestone below and the brown sandstone above the clay bed, because in this larger outcrop they are exposed through a much greater thickness. For example, we may find that the brown sandstone is 180 feet thick and is overlain by a thick bed of distinctive black limestone crowded with fossils.

If, now, we go on to still another outcrop and find here the clay bed is a little thinner than in the last outcrop, if instead of resting on red limestone it now rests on a pink limy sandstone, and if it has a pebbly green sandstone instead of the fine-grained brown sandstone above it, we would be less certain of our correlation, though we would not regard it as impossible.

To check on the correlation, we might go to a lower elevation and study the relations of the sandstone in various outcrops. If we found it gradually varying from limestone to sandy limestone and then to pink sandstone, our correlation would be strengthened.

It would be further confirmed if we walked up the hill and found that the pebbly green sandstone above the clay is overlain by a black limestone full of fossils exactly like those in the limestone that overlay the brown sandstone in the large exposure.

Thus in geologic mapping there is invariably an element of judgment. Some correlations are certain, others are reasonably sure, and for still others there may be a reasonable doubt. As to the doubtful ones, two geologists may disagree, just as two equally qualified physicians sometimes disagree in the diagnosis of identical, but not sufficiently definitive, pathological symptoms.

But nearly all such differences in interpretation have to do with minor features in a stratigraphic succession. Thicker groups of beds commonly exhibit enough peculiarities to lead any two careful observers to identical conclusions.

Geologic Sections

Geologic sections are commonly used along with geologic maps in all economic applications of geology. A geologic section represents the way the rocks would appear on the side of a trench cut vertically into the land surface. It resembles a geologic map in representing the projection of the rock boundaries to a surface, but this surface in a cross section is vertical and not the land surface of the earth.

As an example, let us make a geologic section of the horizontal clay bed just mentioned, taking our data from a geologic map showing the relations of the clay in a deep branching ravine where the beds are well exposed. The completed section appears below the geologic map, together with the construction lines used in drawing it. The geologic section portrays the strata as they would appear on the wall of a trench dug along the line A-B of the geologic map.

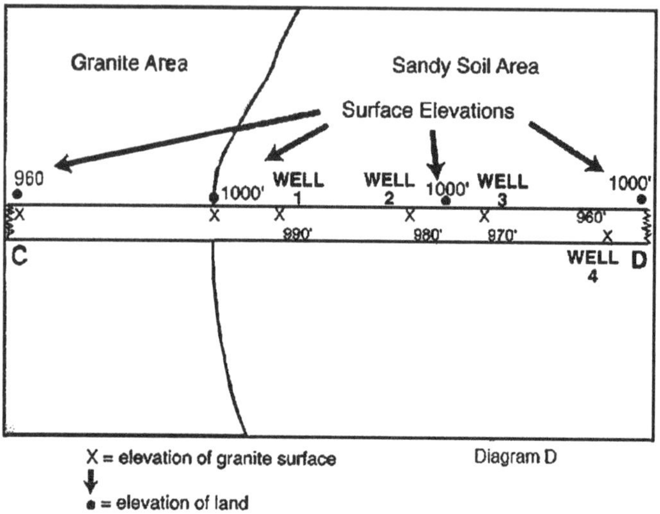

*Fig. **Construction of a Geologic Section from a Geologic Map.***

The section was readily drawn by making use of the contour lines on the map and relating the outcrops to them. First, it was decided to make both the horizontal and vertical scales of the section the same as that of the map, in order to avoid distortion..

Then perpendicular lines were dropped from A and B (by means of a right-angle triangle) to the place below the map chosen for drawing the section. Then a vertical scale showing the range in elevation from 1120 to 1,260 feet above sea level was laid off on the line projected from B. Horizontal lines were drawn across the section, from the point representing the elevation of the respective contours from 1,140 to 1,260. Then a perpendicular was dropped from each point of intersection of line A-13 with a contour line on the map, to the horizontal line on the section representing the appropriate contours. Thus, points were plotted on the section to guide the sketching of the *surface profile*, the irregular line A2 -B2 of the section. Now perpendiculars could be dropped to the profile from the intersection of the map line A-B with each contact of the clay stratum. There are five such intersections shown along the line of section—three on the top of the clay stratum, two on its base. Horizontal lines drawn through the projections of these points on the surface profile complete the cross section and show the relations of the clay bed to the underlying limestone and overlying sandstone.

Many practical uses can be made of geologic sections. If the clay in the bed is suitable for brickmaking, we can determine from the section how much overburden must be removed at any point along the section line so as to get down to the clay. If water is found in wells just above the clay, the section tells us how deeply we must drill at any point along the section in order to strike it. Mine shafts, tunnels, and drill holes all test the validity of our maps and sections and of the postulates underlying their construction.

9

Geomorphology and Landform

LANDFORM REGION

A landform region describes the distinct characteristics of a geographic area, in which some physical characteristics are dominant, almost to the exclusion of others. A landform comprises a geomorphological unit. Landforms are categorized by characteristics such as elevation, slope, orientation, stratification, rock exposure, and soil type. Landforms by name include berms, mounds, hills, cliffs, valleys, and so forth.

Oceans and continents exemplify highest-order landforms. Landform elements are parts of a landform that can be further identified. Landform elements, such as hill-top, shoulder, backslope etc, can be observed on many various geomorphological landforms.

The generic landform elements are: pits, peaks, channels, ridges, passes, pools, planes etc, and can be often extracted from a digital elevation model using some automated or semi-automated techniques. Elementary landforms (segments, facets, relief units) are the smallest homogeneous divisions of the land surface, at the given scale/resolution.

These are areas with relatively homogenuous morphometric properties, bounded by lines of discontinuity.

A plateau or a hill can be observed at various scales ranging from few hundred meters to hundreds of kilometers. Hence, the spatial distribution of landforms is often fuzzy and scale-dependent as is the case for soils and geological strata. A number of factors, ranging from plate tectonics to erosion and deposition can generate and affect landforms. Biological factors can also influence landforms and the work of corals and algae in the formation of coral reefs. Many of the terms are not restricted to refer to features of the planet Earth, and can be used to describe surface features of other planets and similar objects in the Universe.

LANDFORMS OF GEOMORPHOLOGY

Landform

A landform or physical feature in the earth sciences and geology sub-fields, comprises a geomorphological unit, and is largely defined by its surface form and location in the landscape, as part of the terrain, and as such, is typically an element of topography. Landform elements also include seascape and oceanic waterbody interface features such as bays, peninsulas, seas and so forth, including sub-aqueous terrain features such as submersed mountain ranges, volcanoes, and the great ocean basins.

Physical Characteristics

Landforms are categorised by characteristic physical attributes such as elevation, slope, orientation, stratification, rock exposure, and soil type. Gross *physical features or landforms* include intuitive elements such as berms, mounds, hills, ridges, cliffs, valleys, rivers, peninsulas and numerous other structural and size-scaled (i.e. ponds vs. lakes, hills vs. mountains) elements including various kinds of inland and oceanic waterbodies and sub-surface features.

Landform Construction and Destruction

Megageomorphology: In keeping with the regional scale of our examination of major geomorphic provinces we will begin this course by considering tectonic landscapes on the scale of whole mountain ranges and some of the general processes that shape them.

Geomorphologists have also traditionally divided landforms into constructional and destructional categories. Constructional landforms are those that have been or are being built (increasing in mass, height, or area). Examples include tectonic forms such as fault scarps, fault block mountains and valleys (horsts and grabens), rift valleys, monoclines, salt domes, and a variety of volcano landforms. In general, constructional landforms are more common in regions of active tectonism, usually associated with plate margins. Destructional landforms are those that are decreasing in mass, height, or area over time and are primarily the products of erosion and weathering. In truth, all landforms are in part destructional because erosion affects all landforms that lie above sea level on the Earth's surface. However, many landforms are primarily the result of such processes.

More recently, geomorphologists have come to view all landscapes as the products of some combination of three factors:

- tectonics (orogeny, epeirogeny, extension, volcanism)
- erosion
- climate.

None of these factors operates independently-all are in some way functions of the others:

- Tectonics are clearly affected by erosion-removal of crustal mass by erosion leads to isostatic rebound and uplift. In general, about 80% of the height of a landmass removed by erosion will be replaced from below by isostatic readjustment. Some geologists refer to the "pull of erosion" to describe this phenomenon.
- Uplift accelerates erosion, which, as a general rule, increases in intensity with increasing elevation. High elevations also host less vegetation, which armors the land against erosion at low elevations.
- Erosion is a function of climate in many ways. Humid climates as a rule erode faster than arid climates and the effects of erosion on the landscape differ. Humid climates tend to result in the formation of deeply incised valleys. Arid regions erode more evenly. Cold climates result in mountain glaciers that can carve extremely steep topographic features. In fact, mountain glaciers may be Earth's most effective agents of erosion. Extremely frigid climates host large ice sheets that do very little erosion because they are frozen to the bedrock on which they rest.
- Climate itself is greatly affected by tectonism. High landmasses change the balance of heating on the surface and block or redirect the flow of weather systems. As a landmass is elevated its climate cools and may become more humid. This in turn can accelerate erosion. Windward of a mountain range the climate is usually humid as ascending air masses cool and drop their moisture. Leeward of the range is a rain shadow and very arid conditions.

Isostatic mountain building: Climate and erosion can also work together to create uplifted mountain peaks. An arid plateau will support few rivers and tend to erode slowly and uniformly. As material is removed across the plateau isostatic readjustment will cause some uplift and the average land surface will slowly be lowered. If however, the plateau is in moist climate with abundant rivers or glaciers, then abundant, deep valleys can form, removing a large mass of crust, but leaving some crust standing as peaks. Isostatic readjustment will elevate the plateau so that its average mass will be lower than before erosion, but this can result in the peaks being elevated to a higher level than before.

The Three Stages of Mountains: It is probably incorrect to say the mountains are uplifted and then eroded back to sea level. Rather, isostacy, climate and erosion interact to create three stages in the life of a mountain range.

1. Formative Stage-a tectonic event thickens or heats the crust and creates uplift with rates that exceed erosion rates. As the topography rises erosion rates increase.
2. Steady State-erosion rates rise to match uplift rates or uplift slows to match erosion. A steady state results with uplift and erosion rates balanced. Because uplift rates and erosion rates create a negative feedback loop (increasing uplift raises the mountains which intensifies erosion and lowers the mountains) the equilibrium elevation of the mountains may remain stable for millions of years.
3. Decline-if uplift diminishes sufficiently erosion may come to dominate and slowly lower the elevation of the range. This is a slow process because isostatic rebound will replace much of the elevation lost to erosion. By the time a mountain range is worn back to the average level of continental crust, a thickness of crust several times greater than the height of the mountains at their highest can be removed by erosion. For example, it is estimated from mineral studies that the metamorphic rocks of New York City and Connecticut were once buried under about 7 miles of overlying rock. This is not to say that the mountains above them were 7 miles high, only that this thickness of crust has been pulled upward by isostatic readjustment over the last 450 million years.

Major Geomorpho-tectonic Regions

- Coastal Plains-these are extensive regions of flat-lying to gently seaward dipping sediments that build up as a passive continental margin subsides over millions of years. They are areas of very low relief. Generally, coastal plains are absent in tectonically active areas because active uplift of the coastline prohibits their development.
- Orogenic Belts-these are elongate regions that form from collisional tectonics along active continental margins. In presently active orogenic belts the metamorphic rocks may not yet be exposed on the surface-only after extensive erosion of the overlying, intensely deformed sedimentary cover are the metamorphic roots of the mountains exposed. Where they are generated by subduction of oceanic crust beneath the continent they are highly intruded by magmatic plutons.

- Fold and Thrust Belts-marginal to orogenic belts are regions where the sedimentary cover has been folded and thrust faulted by compressional stresses without metamorphism. Low angle thrust faults create blocks of sedimentary crust stacked one on another, suggesting large amounts of shortening of the upper crust (often refered to as "thin-skined tectonics"). Erosion of ancient fold and thrust belts near sea level produces a distinctive ridge and valley topography with surface expressions of underlying syncline and anticline folds.
- Plateaus-these are tectonically elevated regions of undeformed sedimentary rock (the Tibetan Plateau and the Chilean Altiplano are a somewhat different type of plateau-elevated regions within an orogenic belt created by slow erosion due to rain shadow climates). Plateaus developed in in humid regions can be very mountainous due to stream dissection. An extreme form of stream dissection is seen in the canyonlands of the Colorado Plateau, where rivers such as the Colorado have carved mile-deep canyons such as the Grand Canyon.
- Stable Interior/Shield-the interior regions of continents are generally low relief areas with very mild stream dissection. Shield areas in the center of a continent are underlain by ancient, deeply eroded metamorphic and metaigneous rocks. Surrounding the shield regions are areas of the stable interior covered with a thin veneer of nearly horizontal sedimentary strata. These strata are very subtly deformed into broad structurally upwarped regions called domes and downwarped regions called basins which generally do not have any topographic expression.
- Extensional Regions-regions of high mantle heat flow beneath continental crust can cause broad crustal doming and the formation of fault block mountains bounded by grabens and half grabens. Large grabens are also known as rift valleys when they are associated with continental rifting. These features form as the crust is thinned and stretched, creating tensional forces and normal faults. The extensive Basin and Range region of Utah and Nevada in the western US has been generated in the last 17 million years by extension of the western US by as much as 100 miles in the east-west direction. Extensional tectonics are often accompanied by igneous intrusives and surface eruptions as magma works its way upward along faults in the thinned crust.

LAND DEGRADATION AND DESERTIFICATION

Land degradation and desertification are related terms or processes. The accepted UNCED definition is that desertification is land degradation in arid, semi-arid and dry subhumid areas resulting from climatic variation and human activities.

Land degradation threatens economic and physical survival (UNEP, 1999), and could lead to household and national food insecurity in many countries, including the southern African region. Crop yields could be reduced by 50 percent within 40 years if degradation continued at present rates. In South Africa, it is estimated that about 400 million m^3 of soil is lost annually. Key issues to be addressed throughout the region include declining soil fertility, escalating soil erosion, agrochemical pollution and desertification.

Main Types of Erosion and Land Degradation

Land degradation is a composite term, loosely defined as a sustained loss in the quality and the productive capacity of the land. As land degradation progresses, efforts by land users to secure a living become increasingly precarious and uneconomic. The most common indication of land degradation is soil erosion. Subtler but equally important factors include reduction in vegetation cover and changes in vegetation species composition. Stocking and Murnaghan (2000) provide sound practical guidelines on land degradation and its assessment in the field, viewing degradation from the perspective of the land user. How components of land degradation interlink with many other components that influence the quality and productivity of land, including how it is used or misused.

Identification of the direct and indirect causes of land degradation is essential, as any remedial measures designed to rehabilitate land must tackle the root causes of the problem in order for the reversal of land degradation processes to be successful.

Physical Factors

Physical factors always play a role in degradation processes, but their role is less crucial than assumed in erosion hazard mapping. Nevertheless, the topography (slope in particular), the properties of the soil and underlying rock, the vegetation, and climate characteristics (rainfall in particular) are important factors in the acceleration of humaninduced erosion.

Increased runoff and accelerated erosion relate strongly to poor surface conditions, including surface crusts, lack of vegetation, and compaction (decreased infiltration). In addition, the properties of soil horizons and other underlying materials play a major role, especially with respect to gully erosion. Porous materials such as weathered rock (saprolite) and soils with high hydrologic conductivity are conducive to gully erosion as relatively large and rapid subsurface flows may occur, causing collapse of the gully head or sides. Soils with weak structure and friable consistency are vulnerable to erosion, e.g. soils high in illitic or mixed-layer clays. The same applies to soils with high sodicity, inducing a high clay dispersion rate (Solonetz and Planosols).

Climate variability has a profound accelerating effect on erosion and land degradation. Extreme rainfall events aggravate the condition of already degraded land through increased runoff and flooding. Lack of rainfall and resulting drought accelerate desertification processes. Drought acts as a strong catalyst in the initial and progressive degradation of land.

The cause of accelerated erosion is mostly a complex of several factors, as the two main spheres of influence - physical and human - are interrelated and interactive. There is always an element of human influence involved, related to the management of the resources, which may be aggravated by the conditions and characteristics of the environment (climate, soils, geology and landscape). Stable landscapes resistant to erosion normally show less erosion compared with vulnerable environments. Similarly, areas with reliable rainfall in general show less degradation than areas with frequent drought.

Most surveys and studies of land degradation conclude that the primary causes are related to land use, management and socio-economic attitudes. Increase in population is cited as the single most important cause of degradation. However, with land becoming scarce, communities may become more aware of the necessity to improve land management and conservation.

Assessment of Erosion and Land Degradation in the Limpopo River Basin

The Global Assessment of Soil Degradation (GLASOD) by the UNEP is considered the first global assessment of the geographical distribution of human-induced soil degradation (Oldeman, Hakkeling and Sombroek, 1990). Soil degradation is described in terms of: type of erosion and deterioration; cause; degree; rate; and relative extent. The overall status or severity of soil degradation is indicated by a combination of its degree and extent, represented on maps by five classes of severity: none, slight, moderate, high and extreme.

- No degradation, or stable terrain, along the lower northeast part of the Limpopo River Basin in Zimbabwe and Mozambique and in a north-south zone roughly following the Escarpment and associated mountains.
- Slight degradation along the upper Limpopo River Valley, in most of the adjacent southwest catchment in South Africa, and in southeast Zimbabwe. Most of these areas coincide with private farms. Most of the remainder of Mozambique also falls into this class.
- Moderate degradation in northeast Botswana and adjacent Zimbabwe, a north-south zone covering northeast South Africa (including Kruger National Park) and the southern tip of the catchment.
- High degradation in the southwest upper catchment in Botswana and in an area southwest from Pretoria.
- Extreme degradation in three areas in Limpopo Province in South Africa, corresponding with densely populated communal areas (former homelands of Venda and Lebowa).

Apart from the generalized pattern showing several unnaturally shaped units, the validity of the information may be questioned in a number of areas. The strong contrast between the southern part of the Limpopo River Basin and the adjacent Vaal River Basin does not seem realistic. The differences within Botswana do not appear to reflect original information or present status adequately. A substantial part of the South African lowveldt strip has no apparent erosion, notably the area covering Kruger National Park. This same unit should not include Swaziland, where some of the most degraded areas in southern Africa occur.

Hakkeling (1989) described the GLASOD results in more detail for the southern and eastern African regions covering Botswana, Mozambique and Zimbabwe, but not South Africa. This regional map shows a pattern of erosion and degradation in places quite different from that on the integrated global map. In general, the regional map provides a more accurate overview and the country descriptions below draw heavily on this information.

Botswana

The information from the GLASOD regional map (Hakkeling, 1989) indicates that almost all units fall into the overall class of high degradation, except for one area with slight degradation, because of dominant sandy soils. The units most affected by water erosion - both sheet and gully erosion - are found in the northern and southern parts of the Limpopo

River Basin, caused mainly by overexploitation of vegetation and intensive cropping. Wind erosion is the dominant type of degradation in the middle units, caused primarily by overgrazing. Up to 50 percent of all units are considered to have recovered by natural stabilization from earlier degradation.

Reports from the first half of the twentieth century already expressed concern about the state of the land resources. Studies from the second half of the twentieth century indicated that the problem was severe, and that numerous examples of extreme overgrazing could be found. However, it is also claimed that such conclusions are based on flawed data, that the perception of range condition and management is biased in favour of western models, and that most of the examples of degradation are temporary conditions that are natural to a variable savannah ecosystem.

Evidence is based mainly on examples of: bush encroachment; decrease in general grass density and in numbers of the more nutritious and palatable species; and of an increase in patches of bare soil. The consensus is that rangeland degradation is occurring, but there is disagreement as to its extent, severity and reversibility. Abel and Blaikie (1989) argue that degradation should not be measured by using indicators such as short-term vegetation changes. Instead, changes in the environment should be judged according to their degree of irreversibility over longer periods. The assessment of detrimental changes must also take into account estimates of secondary production interest of the users, for which reason the productivity of communal grazing land may be higher than that of commercial ranches (De Ridder and Wagenaar, 1984).

The description of erosion and land degradation within the Limpopo River Basin (hardveldt) of Botswana is confined mostly to local observations, and although overviews have been produced (Arntzen and Veenendaal, 1986; Dahlberg, 1994), a comprehensive inventory is lacking. Ringrose and Matheson (1986) reported an increase in desertification manifested as decreased vegetation, increased erosion and reduction of soil water retention as a result of loss of soil organic matter induced by overgrazing and fuelwood collection. In this regard, the BRIMP also generates various degradation maps and datasets, based on ground monitoring and the interpretation of Landsat images specifically aimed at addressing desertification questions.

Occurrence of severe sheet and gully erosion is reported from several sites in east Botswana near Serowe and Kalamare, but erosion is generally estimated to be slight or moderate. Most of the erosion is associated with sloping land, including the footslopes of hills. The flat and slightly undulating parts of the plains show less evidence. Wind erosion is reported from areas with a bare surface, especially from fallow arable land, but the severity is difficult to estimate. This effect is most pronounced after periods of drought and reduction of the protective vegetation cover.

Mozambique

The information from the GLASOD regional map indicates that the northernmost units of the Limpopo River Basin have no livestock because of tsetse fly, hence, no erosion is described. Most other units have the overall class of slight degradation, except for two units with moderate degradation. High degradation is reported in the Changane Valley owing to crusting and sealing. Most of the degradation is caused by wind erosion - with some nutrient losses - but a variety of causes have been observed: overgrazing in the Lebombo Hills, intensive cropping along the coast, and salinization in the irrigated areas near Xai-Xai in the Limpopo River floodplain. Moreover, natural stabilization seems a common process in Mozambique.

Assessment of erosion risk in Mozambique was first undertaken on a national scale by FAO (1985) when Reddy and Mussage compiled a first approximation of an erosive capacity index. The low rainfall areas of Gaza Province were classified as a low erosion-hazard zone. High erosion-risk zones included the coastal areas of Gaza Province. Population density in the coastal belt has raised concern about dune vegetation, mangroves and coral reefs. While none of these ecotypes could yet be considered critically threatened in Mozambique, local areas of degradation have been identified and the government is anxious to take remedial action before the problems become more severe.

The various studies and surveys reviewed indicate that major problems in the development of the Limpopo Valley soils are related to salinity and alkalinity. These problems arise from the geological structure, which has very saline marine sediments under the recent alluvial sediments (primary salinization). Furthermore, mismanagement of the irrigation water supplies and of the drainage systems has raised groundwater tables and led to the efflorescence of salts at the ground surface (secondary salinization).

South Africa

The GLASOD map of Africa (Oldeman, Hakkeling and Sombroek, 1990) generally indicates less soil degradation in the Limpopo River Basin as compared with the rest of South Africa. Most of Limpopo Province shows none to slight degradation, with moderate degradation in the eastern parts, all caused by water erosion, and primarily caused by overgrazing. Moderate to high degradation resulting from pollution and acidification occurs in the south of the catchment (Pretoria-Johannesburg-Witbank area). The most severe soil degradation caused by overgrazing occurs in the former homeland areas in the central Limpopo Province.

Hoffman and Todd (1999) have provided an overview of erosion and land degradation in South Africa, in a first-phase review for the development of South Africa's national action programme to combat desertification. A number of general conclusions from this report were presented, which are also relevant to the Limpopo River Basin. The focus of land degradation has historically been on the degradation of vegetation, in particular of the rangelands. The study suggests that equal attention should be paid to degradation of vegetation, soil and water resources.

Earlier inventories of soil and land degradation have concentrated on erosion hazard with relatively little attention given to the influence of land use practices. The present view is that both land use and land tenure exercises have a significant influence on land degradation.

Hoffman and Todd (1999) state that soil and vegetation degradation is perceived as being significantly greater in communal areas as compared with commercial areas, by at least a factor of two. However, specific forms of vegetation degradation are more of a problem in commercial areas, such as change in species composition, alien plant invasion, and encroachment of indigenous woody species.

It is suggested that areas with steep slopes, low rainfall and higher temperatures are significantly more eroded. Climate change in the last century may have had an impact on the intensity of erosion, but this needs to be further studied, in particular in order to define the relationship between drought and erosion.

The government policy of land allocation has had a major effect on land use and land degradation. In the commercial farming areas, this has been partially conducive to sustainable land use, but not so in the communal areas, where practices of crop production, communal grazing and use of the vegetation have led to accelerated erosion and degradation. The understanding in South Africa of communal land use and production systems is poorly developed, as is its relationship with land degradation (Hoffman and Todd, 1999).

In an overview on soil degradation in South Africa, a report by FAO (1998a) states that data on degradation are incomplete and fragmented, with little information on spatial distribution and distinction between natural and human-induced erosion. Relevant land degradation information is available from a number of other sources (Newby and Wessels, 1997; Van Zyl, 1997; Scotney, 1995; Laker, 1994).

Nationally, it is estimated that water erosion affects 6.1 million ha, or more than 40 percent of the total of 14.6 million ha of cultivated soil in South Africa. Of this, 15 percent is seriously affected, 37 percent moderately affected and the remaining 48 percent slightly affected. Laker (1994) estimates that about 20 percent of all topsoil has been lost since the beginning of the twentieth century. Average annual soil loss is estimated at 2.5 tonnes/ha, equal to about 300 million tons. Wind erosion affects even more cultivated land, an estimated 10.9 million ha, or more than 70 percent. Of the affected cultivated land, 7 percent is seriously eroded, 29 percent moderately and 64 percent slightly. Compaction affects about 2 million ha, or about 15 percent of all cultivated land. Surface sealing is also a widespread and a serious problem. It is estimated that a total of 15 percent or more of the 1.2 million ha of irrigated land is moderately to severely affected by salinization and/or waterlogging. In addition, large areas are affected by pollution, acidification and fertility losses. Of the national grazing land about 3 million ha are rendered worthless for grazing as a result of encroachment by undesirable species.

The occurrence of soil and vegetation degradation combined in a single index per magisterial district. The provinces within the Limpopo River Basin are described as belonging to the most eroded and degraded parts of the country (which contradicts the GLASOD result). However, these estimates are based on perceptions from agricultural extension and resource conservation officers, and not on actual observations.

According to Newby and Wessels (1997), the estimated extent of degradation of vegetative cover is 25 percent for North West Province (poor to very poor condition), 87 percent for Limpopo Province (poor to critical condition) and 50 percent for Mpumalanga Province (poor to critical condition). Serious wind and water erosion is reported from North West Province, respectively 53 and 36 percent of all area affected. The mining industry - predominantly coal mining - has occupied and severely polluted 50 percent of all high potential arable land in Mpumalanga. However, this statement is contrary to the experience and results of other researchers.

Wessels *et al.* (2001) mapped the conservation status of natural vegetation and soils in Mpumalanga Province, including the present land cover, rangeland condition, alien vegetation and bush encroachment. Some of their main findings. One-third or more of both the grassland and savannah biomes in the province is stated to be in poor to very poor condition. Visible soil erosion damage is stated to occur over less than one-third of the province.

Zimbabwe

Information from the GLASOD regional map indicates overall strong degradation in the communal areas in the northern and western part of the Limpopo River Basin adjacent to Botswana, with severe sheet and gully erosion caused by deforestation and intensive cropping. No erosion is described for the areas with commercial ranching as the dominant land use, and for national parks.

The communal grazing area in the east has moderate soil degradation. As in Botswana, most units are considered to have stabilized in a natural way from earlier degradation.

The history of land degradation in Zimbabwe and the different views about it are quite similar to Botswana. Official interventions based on reports citing the occurrence of severe soil erosion have largely failed. The increase in crop and animal production in the 1980s is considered sufficient evidence that there is no substantial decline in the natural resource base. However, this may be a preliminary conclusion as the 1990s have shown a general decline.

Campbell *et al.* (1990) claim that there is sufficient evidence to indicate that cattle are linked to major environmental changes, but they also conclude that the issue of whether communal grazing practices cause degradation cannot be answered without additional data, in particular a spatial analysis.

Scoones (1989b, 1992b, 1993) finds that disregard of the heterogeneity of the environment is one of the main reasons for conflicting perceptions of environmental change. Areas will differ on the extent of degradation and the effects of the changes on secondary production. Scoones found that in most years the farmers are avoiding irreversible damage, in terms of productivity, by using strategies of herd mobility and local ecological knowledge. In drought years, migration between different savannah types is employed on a regional scale. Certain elements of the landscape or ecosystem, such as riverine strips, vleis (wet bottomlands) and drainage lines, are vital in providing fodder in critical periods.

The extent of soil erosion in Zimbabwe was mapped nationally using air photography. Only 5 percent of the country was badly eroded, in comparison with communal areas where badly eroded land comprised about 10 percent of the land area. Little difference was observed between cropland and grazing land. This is somewhat surprising as most arable land in Zimbabwe seems adequately protected by contour ridges. In other countries in southern Africa, erosion is always reported to be excessively higher on communal grazing land, e.g. in Swaziland (Remmelzwaal and McDermott, 1997).

Elwell (1980) calculated annual soil losses on cropland caused by sheet erosion followed by national surveys. Grohs (1994) re-interpreted Elwell's results according to administrative areas. The highest soil losses of more than 100 tonnes/ha were recorded in the north and east of the country. The lowest values of less than 5 tonnes/ha were found in the semi-arid south, within the Limpopo River Basin. On the basis of these findings, an average annual soil loss of about 55 million tonnes was calculated.

10

Geologic Features in Structural Geology

Structural geology deals with the geometric relationships of rocks and geologic features in general. The scope of structural geology is vast, ranging in size from submicroscopic lattice defects in crystals to mountain belts and plate boundaries.

Structures may be divided into two broad classes: the primary structures that were acquired in the genesis of a rock mass and the secondary structures that result from later deformation of the primary structures. Most layered rocks (sedimentary rocks, some lava flows, and pyroclastic deposits) were deposited initially as nearly horizontal layers. Rocks that were initially horizontal may be deformed later by folding and may be displaced along fractures. If displacement has occurred and the rocks on the two sides of the fracture have moved in opposite directions from each other, the fracture is termed a fault; if displacement has not occurred, the fracture is called a joint. It is clear that faults and joints are secondary structures; i.e., their relative age is younger than the rocks that they intersect, but their age may be only slightly younger. Many joints in igneous rocks, for example, were produced by contraction when the rocks cooled. On the other hand, some fractures in rocks, including igneous rocks, are related to weathering processes and expansion associated with removal of overlying load. These will have been produced long after the rocks were formed. The faults and joints referred to above are brittle structures that form as discrete fractures within otherwise undeformed rocks in cool upper levels of the crust. In contrast, ductile structures result from permanent changes throughout a wide body of deformed rock at higher temperatures and pressures in deeper crustal levels. Such structures include folds and cleavage in slate belts, foliation in gneisses, and mineral lineation in metamorphic rocks.

The methods of structural geology are diverse. At the smallest scale, lattice defects and dislocations in crystals can be studied in images enlarged several thousand times with transmission electron microscopes. Many structures can be examined microscopically, using the same general techniques employed in petrology, in which sections of rock mounted on glass slides are ground very thin and are then examined by transmitted light with polarizing microscopes. Of course, some structures can be studied in hand specimens, which were preferably oriented when collected in the field.

On a large scale, the techniques of field geology are employed. These include the preparation of geologic maps that show the areal distribution of geologic units selected for representation on the map. They also include the plotting of the orientation of such structural features as faults, joints, cleavage, small folds, and the attitude of beds with respect to three-dimensional space.

A common objective is to interpret the structure at some depth below the surface. It is possible to infer with some degree of accuracy the structure beneath the surface by using information available at the surface. If geologic information from drill holes or mine openings is available, however, the configuration of rocks in the subsurface commonly may be interpreted with much greater assurance as compared with interpretations involving projection to depth based largely on information obtained at the surface. Vertical graphic sections are widely used to show the configuration of rocks beneath the surface.

Balancing cross sections is an important technique in thrust belts. The lengths of individual thrust slices are added up and the total restored length is compared with the present length of the section and thus the percentage of shortening across the thrust belt can be calculated. In addition, contourmaps that portray the elevation of particular layers with respect to sea level or some other datum are widely used, as are contour maps that represent thickness variations.

Strain analysis is another important technique of structural geology. Strain is change in shape; for example, by measuring the elliptical shape of deformed ooliths or concretions that must originally have been circular, it is possible to make a quantitative analysis of the strain patterns in deformed sediments. Other useful kinds of strain markers are deformed fossils, conglomerate pebbles, and vesicles. A long-term aim of such analysis is to determine the strain variations across entire segments of mountain belts. This information is expected to help geologists understand the mechanisms involved in the formation of such belts.

A combination of structural and geophysical methods are generally used to conduct field studies of the large-scale tectonic features mentioned below. Field work enables the mapping of the structures at the surface, and geophysical methods involving the study of seismic activity, magnetism, and gravity make possible the determination of the subsurface structures. The processes that affect geologic structures rarely can be observed directly. The nature of thedeforming forces and the manner in which the Earth's materials deform under stress can be studied experimentally and theoretically, however, thus providing insight into the forces of nature.

One form of laboratory experimentation involves the deformation of small, cylindrical specimens of rocks under very high pressures. Other experimental methods include the use of scale models of folds and faults consisting of soft, layered materials, in which the objective is to simulate the behaviour of real strata that have undergone deformation on a larger scale over much longer time.

Some experiments measure the main physical variables that control rock deformation—namely, temperature, pressure, deformation rate, and the presence of fluids such as water. These variables are responsible for changing the rheology of rocks from rigid and brittle at or near the Earth's surface to weak and ductile at great depths. Thus, experimental studies aim to define the conditions under which deformation occurs throughout the Earth's crust.

STRUCTURAL GEOLOGY

Structural geology is the study of the three-dimensional distribution of rock units with respect to their deformational histories. The primary goal of structural geology is to use measurements of present-day rock geometries to uncover information about the history of deformation (strain) in the rocks, and ultimately, to understand the stress field that resulted in the observed strain and geometries. This understanding of the dynamics of the stress field can be linked to important events in the regional geologic past; a common goal is to understand the structural evolution of a particular area with respect to regionally widespread patterns of rock deformation (e.g., mountain building, rifting) due to plate tectonics.

Use and Importance

The study of geologic structures has been of prime importance in economic geology, both petroleum geology and mining geology. Folded and faulted rock strata commonly form traps for the accumulation and concentration of fluids such as petroleum and natural gas. Faulted and structurally complex areas are notable as permeable zones for hydrothermal fluids and the resulting concentration areas for base and precious metal ore deposits. Veins of minerals containing various metals commonly occupy faults and fractures in structurally complex areas. These structurally fractured and faulted zones often occur in association with intrusive igneous rocks. They often also occur around geologic reef complexes and collapse features such as ancient sinkholes. Deposits of gold, silver, copper, lead, zinc, and other metals, are commonly located in structurally complex areas.

Structural geology is a critical part of engineering geology, which is concerned with the physical and mechanical properties of natural rocks. Structural fabrics and defects such as faults, folds, foliations and joints are internal weaknesses of rocks which may affect the stability of human engineered structures such as dams, road cuts, open pit mines and underground mines or road tunnels.

Geotechnical risk, including earthquake risk can only be investigated by inspecting a combination of structural geology and geomorphology. In addition areas of karst landscapes which are underlain by underground caverns and potential sinkholes or collapse features are of importance for these scientists. In addition, areas of steep slopes are potential collapse or landslide hazards.

Environmental geologists and hydrogeologists or hydrologists need to understand structural geology because structures are sites of groundwater flow and penetration, which may affect, for instance, seepage of toxic substances from waste dumps, or seepage of salty water into aquifers.

Plate tectonics is a theory developed during the 1960s which describes the movement of continents by way of the separation and collision of crustal plates. It is in a sense structural geology on a planet scale, and is used throughout structural geology as a framework to analyze and understand global, regional, and local scale features.

Methods

Structural geologists use a variety of methods to (first) measure rock geometries, (second) reconstruct their deformational histories, and (third) calculate the stress field that resulted in that deformation.

Geometries

Primary data sets for structural geology are collected in the field. Structural geologists measure a variety of planar features (bedding planes, foliation planes, fold axial planes, fault planes, and joints), and linear features (stretching lineations, in which minerals are ductily extended; fold axes; and intersection lineations, the trace of a planar feature on another planar surface).

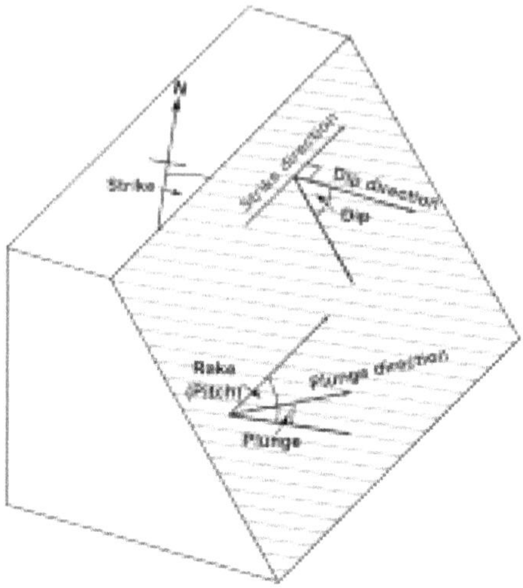

Fig. Illustration of measurement conventions for planar and linear structures

Measurement Conventions

The inclination of a planar structure in geology is measured by *strike and dip*. The strike is the line of intersection between the planar feature and a horizontal plane, taken according to the right hand convention, and the dip is the magnitude of the inclination, below horizontal, at right angles to strike. For example; striking 25 degrees East of North, dipping 45 degrees Southeast, recorded as N25E,45SE.

Alternatively, dip and dip direction may be used as this is absolute. Dip direction is measured in 360 degrees, generally clockwise from North. For example, a dip of 45 degrees towards 115 degrees azimuth, recorded as 45/115. Note that this is the same as above. The term *hade* is occasionally used and is the deviation of a plane from vertical i.e. (90°-dip). Fold axis plunge is measured in dip and dip direction (strictly, plunge and azimuth of plunge). The orientation of a fold axial plane is measured in strike and dip or dip and dip direction.

Lineations are measured in terms of dip and dip direction, if possible. Often lineations occur expressed on a planar surface and can be difficult to measure directly. In this case, the lineation may be measured from the horizontal as a *rake* or *pitch* upon the surface.

Rake is measured by placing a protractor flat on the planar surface, with the flat edge horizontal and measuring the angle of the lineation clockwise from horizontal. The orientation of the lineation can then be calculated from the rake and strike-dip information of the plane it was measured from, using a stereographic projection.

If a fault has lineations formed by movement on the plane, e.g.; slickensides, this is recorded as a lineation, with a rake, and annotated as to the indication of throw on the fault.

Generally it is easier to record strike and dip information of planar structures in dip/dip direction format as this will match all the other structural information you may be recording about folds, lineations, etc., although there is an advantage to using different formats that discriminate between planar and linear data.

Plane, Fabric, Fold and Deformation Conventions

The convention for analysing structural geology is to identify the planar structures, often called *planar fabrics* because this implies a textural formation, thelinear structures and, from analysis of these, unravel deformations. Planar structures are named according to their order of formation, with original sedimentary layering the lowest at S0. Often it is impossible to identify S0 in highly deformed rocks, so numbering may be started at an arbitrary number or given a letter (S_A, for instance). In cases where there is a bedding-planefoliation caused by burial metamorphism or diagenesis this may be enumerated as S0a.

If there are folds, these are numbered as F_1, F_2, etc. Generally the axial plane foliation or cleavage of a fold is created during folding, and the number convention should match. For example, an F_2 fold should have an S_2 axial foliation. Deformations are numbered according to their order of formation with the letter D denoting a deformation event. For example D_1, D_2, D_3. Folds and foliations, because they are formed by deformation events, should correlate with these events. For example an F_2 fold, with an S_2 axial plane foliation would be the result of a D_2 deformation.

Metamorphic events may span multiple deformations. Sometimes it is useful to identify them similarly to the structural features for which they are responsible, e.g.; M_2. This may be possible by observing porphyroblast formation in cleavages of known deformation age, by identifying metamorphic mineral assemblages created by different events, or via geochronology.

Intersection lineations in rocks, as they are the product of the intersection of two planar structures, are named according to the two planar structures from which they are formed. For instance, the intersection lineation of a S_1 cleavage and bedding is the $L_{1\text{-}0}$ intersection lineation (also known as the cleavage-bedding lineation).

Stretching lineations may be difficult to quantify, especially in highly stretched ductile rocks where minimal foliation information is preserved. Where possible, when correlated with deformations (as few are formed in folds, and many are not strictly associated with planar foliations), they may be identified similar to planar surfaces and folds, e.g.; L_1, L_2. For convenience some geologists prefer to annotate them with a subscript S, for example L_{s1} to differentiate them from intersection lineations, though this is generally redundant.

STRUCTURAL GEOLOGY AND ATTITUDE OF ROCK LAYERS

No attempt was made to construct a map showing primarily the attitude of rock layers. To do so would be a simple engineering project, as there are numerous good "key beds," and the intervals between prominent limestone beds are remarkably constant. Geologists of several oil-producing companies have worked in the two counties and it is very probable that accurate information concerning geologic structure is available to those persons most interested in having it. For that reason the structure is discussed in only a general way and from the viewpoint of those features that were obvious to me when I made the areal geologic map.

The consolidated rocks are structurally included in the Prairie Plains monocline, i.e., beds dipping westward or slightly north of westward from the Ozark plateau. The general inclination is about 30 feet per mile. This dipping of beds can be plainly observed; for example, the Cottonwood limestone high on the hills at Manhattan disappears under the alluvial filling of Kansas River valley a short distance west of Eureka Lake (Odd Fellows' Home), and the Fort Riley limestone descends nearer and nearer the floor of the river valley as one goes westward from Junction City along Smoky Hill river. Other beds of stratified rock are approximately parallel to those just cited. There are minor distortions in the position of the layers of rock, and hence in places the general westward dip is steepened or locally the beds dip in some other direction. In southeastern Riley County there is a structural dome or anticline and the valley of Deep Creek is superposed upon it. This upwarping of strata is nearly in superposition on the buried Nemaha Mountains and is a part of the anticlinal structure that crosses the state above the buried mountains. From the distribution of outcrops in relation to topography it seems that the crest of the fold in Deep Creek valley is probably a few miles southeast of Zeandale, but the alluvium of Deep Creek valley has obscured the bedrock rock over an extensive area, so the statement must be qualified. Exposures in the Flint Hills west of Zeandale show a steep westward dip and the rocks there lie upon the west limb of the Deep Creek fold. Plate, 4A shows this condition.

The buried mountains occupy a band a few miles wide crossing Kansas from Cowley County to Nemaha County, along which the surface of the very,ancient crystalline rocks, above which were deposited younger stratified sediments, is much higher and hence much nearer the present land surface than it is on either side. The mountains were described by Moore and Haynes (1917, p. 173) and are well known to all students of the subsurface geology of the Midcontinent region. The Nemaha Mountains extend well into Oklahoma and into Nebraska. The mountains were named Nemaha because they are not many hundreds of feet below the surface in Nemaha County, Kansas. Many geologists have discussed the probable origin of the superposed folds in relation to the buried topography. It is very probable that the flexures in rock above the buried hills and valleys are at least partly due to differential compaction and settling of the younger rock layers above the rigid crystalline rocks following the uneven surface below. It is likewise probable that the distortion is partly due to tangential compressive forces acting on the rocks. Bass (1929, pp. 126-128) has discussed the origin of structural features in southern Kansas, and inasmuch as structural features above the buried hills in Riley County are very similar, his discussions should be consulted by those who are concerned with the origin of the folds in the area of this report.

There is a noticeable arching of beds in the vicinity of Winkler, in northern Riley County in the southeast part of T. 6 S., R. 6 E. Near the town, exposures of Florence limestone in the Barneston formation show that the Wreford limestone beds must lie well below the surface of the alluvium of the flood plain of Fancy Creek; but only a short distance westward the Wreford formation lies well above the almost flat flood plain, and still farther westward it dips below the stream again. The same reversal of dips can be plainly seen in the bluffs forming the walls of Fancy Creek valley and even more plainly along streams tributary to Fancy Creek on the north.

As the base of the Wreford formation is indicated on the map showing areal geology, the arch near Winkler is indicated on the map by an inlier of the Wreford formation along Fancy Creek. The fold described here is a part of the Abilene anticline, which extends from a point near Kingman in Kingman County in the south-central part of Kansas northward into

Nebraska. In Nebraska it is commonly called the Barneston arch. Probably it is no more acute near Winkler than elsewhere along its axis, although elsewhere the flat topography may not reveal the arching so well.

FOLDING MECHANISMS

Folding of rocks must balance the deformation of layers with the conservation of volume in a rock mass. This occurs by several mechanisms.

Flexural Slip

Flexural slip allows folding by creating layer-parallel slip between the layers of the folded strata, which, altogether, result in deformation. The best analogy is bending a phone book, where volume preservation is accommodated by slip between the pages of the book.

Buckling

Typically, folding is thought to occur by simple buckling of a planar surface and its confining volume. The volume change is accommodated by *layer parallel shortening* the volume, which grows in *thickness*. Folding under this mechanism is typically of the similar fold style, as thinned limbs are shortened horizontally and thickened hinges do so vertically.

Mass Displacement

If the folding deformation cannot be accommodated by flexural slip or volume-change shortening (buckling), the rocks are generally removed from the path of the stress. This is achieved by pressure dissolution, a form of metamorphic process, in which rocks shorten by dissolving constituents in areas of high strain and redepositing them in areas of lower strain. Folds created in this way include examples inmigmatites, and areas with a strong axial planar cleavage.

Folding in the South Tien Shan

The Hercynian fold system of the South Tien Shan is well known for the wide development of different morphological types of folding both in the Hercynian and Alpine rock complexes. The South Tien Shan is a standard area, the long study of which resulted in the formulation of a main theory for the Alpine orogenesis of old fold mountains [*Shultz,* 1937], for deep crustal faults (papers by A. V. Peive and A. Suvorov), for the facies and formations of sedimentary and igneous rocks, and the views were advanced for the geology and mechanics of shears and nappes.

The territory is divided into structural zones, where the main fold structures were mapped: anticlinoria and synclinoria. However, the systematic studies of the Paleozoic folding were initiated there as late as the seventies of the 20th century. Yet, the problems of the structural relations between the fold zones, nappes, and deep faults remained unsolved.

At the end of the 1980s, views were advanced that the fold deformations that produced the structural pattern of this zone had operated during the Alpine period of the geological history [*Yablonskaya,* 1989], rather than during the Hercynian time, as believed earlier. No consensus has been achieved on the dating of this fold structure. At the same time this problem remains to be urgent, because many of the modern reconstructions are based on different views on the structure of the South Tien Shan fold system. In particular, the highly debatable item is the dominant role of the horizontal compression exerted on the Tien Shan from the Hindustan protrusion of the Gondwanaland and from the lithospheric blocks of the Pamirs and Afgan-Tajik mountain massifs during the formation of the recent, and possibly, of the Paleozoic fold-and-nappe structure of this mobile system. In my study I investigated, with a varying degree of detail, the structural geology of several areas in the South Tien Shan region. The results of these studies allowed me to return to some unsolved problems of the South Tien Shan, formulated in the 1970s and 1980s. In my interpretation of the field data I used the methods of the qualitative and quantitative descriptions of folds and fold-related faults.

This work revealed the great morphologic heterogeneity of the folds in the area of this mountain system. The disharmonic attitudes of size-varying folds in the sedimentary and volcanic rock sequences, as well as the structural relationships among the folds, faults, and nappes, indicated the complex multifactor character of the folding process. The comparison of the folding intensity in the Paleozoic and Meso-Cenozoic rocks allowed me to trace the evolution of the holomorphic folding pattern formation in time.

Methods of Data Collection and Interpretation : In order to collect information on the real forms of folds and faults, special-purpose field surveys were carried out in the seventies and eighties of the last century across the strike of the fold system. The documented materials included the description of the structural features of varying-size folds, plication, cleavage, boudinage and the overlapping of the folds of two or more generations in the sediments and, especially, in the metamorphic rocks. Where large folds could not be traced or were present in fragments, the monoclinal rock packages were examined. Particular attention was given to the exact location of the bottoms and tops of the layers. All large and small breaks and faults were mapped with a special attention given to the zones of regional faults. All data collected for the faults and folds were plotted in geological sections of 1:10 000 or 1:25 000 scales. These sections were plotted across the strikes of the fold zones at equal distances from one another. As a result, the study areas were covered, more or less equally, by structural observations.

The analysis of the cross-sections does not give a complete idea of the region's structure. One can only locate and trace, along the strike, the zones of different types of disturbances. Whereas the relations among the dislocations of different orders and the zonal pattern of folding could be understood only in the course of areal studies. The work of this kind was done in the most complicated areas and in the areas where the study profiles were comparatively rare [*Rogozhin,* 1977].

The main purpose of this study is the detailed classification of folding into morphological types. It had been believed conventionally that in terms of its morphology the Paleozoic folding pattern in the South Tien Shan region was uniform and linear, that is, holomorphic, whereas the results of field observations suggest the substantial heterogeneity of the folding pattern.

To be more objective in the location and tracing of the longitudinal and transverse heterogeneities of any fold structure, we used two qualitative methods of estimating the complexity of the folds. One of them was the formalized method of the quantitative estimation of morphological folding complexity (MFC) [*Rogozhin,* 1986]. This method had been devised and offered for analyzing series of detailed geological sections crossing fold areas across their strikes. Such estimations are performed in previously located cross-section segments, commensurable in size, or structural domains, showing a network of morphological indications, the sum total of which is necessary and sufficient for ranking the mapped folding types with the known morphological varieties. The folding morphology is comparatively uniform in each domain. The boundaries between the domains are drawn in the places of notable changes in the folding morphology or along the zones of large faults.

We used the following morphologic indications of fold deformations: (1) an angle between the fold limbs, (2) a degree of similitude between anticlinal and synclinal forms, (3) the presence of cross-cutting cleavage, (4) a ratio between the fold curve and limb widths, (5) the dip angle of fold axial surfaces, (6) a ratio between the numbers of larger and smaller folds, and (7) the position of a fine folding plane. The holomorphic (MFC = 1.0) and intermediate (MFC = 0.1) folding types, characteristic of the central and peripheral zones of the fold systems are usually found, as a result of using this analytical approach, to be the extreme members of the gradual series of smaller morphological forms. Each of the indications was estimated to be equal to 0.1 or 0.2 for highly significant indications. As a result, each structural domain in each cross-section has a series of estimates for each indication and a sum total can be estimated.

In my study I usually analyzed the forms of small folds, which can be observed in the field. These are folds of the third and fourth orders of magnitude. Four classes of folds were located confidently in the South Tien Shan region. The first type includes the folds whose widths and heights measured a few kilometers and lengths, dozens of kilometers. In terms of the third order of magnitude, these values are hundreds of meters and kilometers, respectively. Fourth-order folds were observed in outcrops, because their sizes measure meters and dozens of meters (and may be as large as a hundred of meters).

The MFC estimates for the domains located in each geological section allow one to trace and estimate quantitatively the heterogeneity of the forms of small folds over an area by way of interpolating the values obtained for all geologic sections available.

The resulting MFC values can be compared with the results of the earlier used methods of the qualitative estimation of folding intensity, which are more usual for many geologists [*Sholpo,* 1978]. The highest MFC values (0.75-1.0) are known for the zones of most compressed isoclinal folds, the intermediate MFC values (0.6-0.7), for the zones of folds with less intense curves, comparatively low MFC values (0.3-0.55), for the zones of arcuate or round forms, and the lowest MFC values (0.1-0.25), for carinate and box-shaped folds. Another quantitative parameter used to estimate the compression degree of the folded rocks was the so-called coefficient of the excessive length of the layers (l), which is derived by calculating the ratio of the curvimeter-measured length of the layers of folded rocks measured using a curvature meter in the cross section to the horizontal length of the measured rock segment [*Vikhert,* 1972]. This characteristics reflects the degree of the horizontal reduction of the sedimentary rock basis in the course of folding, that is, the excessive length of the layers. This estimate allows one to compare the cumulative fold deformation, measured in the Paleozoic rocks, and the recent deformation, measured in the rock sequences of Paleozoic age, with the recent deformation estimated only in the Mesozoic and Cenozoic deposits.

The morphology of the co-folding faults was studied using the conventional methods employed in structural geology [*Belousov,* 1975, 1985]. The inclined faults recorded in each section were classified in terms of their morphology as normal faults, reverse faults, or thrust faults. Using the amounts of the relative displacements of the rock layers of the same age, the absolute amounts of the vertical and horizontal displacements in the fault sides were found and the faults were classified as normal or upthrust faults, and the relative amounts of compression and extension were calculated in per cent relative to the total length of the section.

Displacements along the faults were estimated in large fold structures, such as anticlinoria and synclinoria, this allowing one to characterize the latter not only in terms of the heterogeneity of the plicated rocks, but also in terms of the contribution of the faults into the general deformation pattern.

Morphological Heterogeneity of the Fault and Fold Structure : The Turkestan and Alai structural features have been studied in most detail. The fold zones are characterized there by a complex linear folding style and by the development of huge anticlinoria and synclinoria against the background of widely developed pre-folding nappes.

The Zeravshan-Gissar zone was studied along individual cross-section lines. Yet, the detailed structural studies done there by *Leonov* [1985, 1990] allowed us to have an idea of its fold-and-nappe structure, with numerous imbricate nappes which do not fit one system of nappes [*Leonov,* 1990].

Discarded from our study were the South Gissar and Baisun tectonic zones, because the former is almost wholly occupied by the granites of the Gissar Pluton, and the latter is poorly exposed, this precluding the systematic study of the folding morphology. Finally, the Baubashata mountain area, located northeast of the Fergana Basin, is characterized by only one cross-section, this demanding the discussion of this cross section only on the condition of comparing it with the data from the better known zones.

Our study of the above mentioned areas in the South Tien Shan region allowed us to get a general idea of its structure in the central segment of the fold zone, the main structural elements of which are tectonic nappes, large anticlinoria and synclinoria, as well as numerous smaller folds, and a system of regional and local faults.

The formation of the Hercynian tectonic nappes preceded the main folding epoch. The packages of the nappes (sheets) experienced folding, like the stratigraphic rock sequences, and participated in the formation of synclinoria and, to a lesser extent, of anticlinoria. Associated with the movements of the tectonic nappes, was the formation of early drag folds, encountered in the allochthonous sheets. These folds are best developed and are most numerous in the frontal parts of the latter and are scarce in their backs, where the beds experience subhorizontal stretching.

The tectonic zones that had been mapped earlier in the area of the South Tien Shan show substantial differences in the composition and thickness of the Middle Paleozoic rocks and, hence, in the tectonic environment in which these rocks had deposited.

Widely developed and bordering one another are the zones of "shelf" (carbonate-terrigenous) sedimentation, "island-arc" (volcanic-type) evolution, and "paleooceanic" (condensed-type siliceous rocks and ophiolites) of the Paleozoic history. The specific feature, common for the Paleozoic rocks in all zones, is the fact that the clay-shale rock sequences of Silurian and, partly, of Early Paleozoic age, composing the bottoms of all types of rock sequences, are stratigraphically higher replaced everywhere by more compact rocks, such as the Middle Paleozoic volcanic, siliceous, or carbonate rocks. Therefore, all types of rock sequences show a density inversion.

The tectonic nappes are composed of Middle Paleozoic volcanic and siliceous ophiolitic rocks. They are most abundant in the northern zones of the South Tien Shan region, namely, in the Karachatyr, Turkestan-Alai, and Baubashata zones at the boundary with the Fergana-Kyzylkum Median Massif and are almost totally absent in the Zeravshan-Turkestan and Zeravshan-Gissar axial zones of the system. It is remarkable that overthrusts are equally well developed in the areas of the latest intensive folding, and in the cases of its relatively mild manifestations, this suggesting the independent developments of the folds and overthrusts. At the same time there is an obvious relationship between the regional faults and large fold zones: the faults divide many anticlinoria and synclinoria, some synclinoria having been formed along large fault zones.

Each of the largest fold structures shows an abrupt repeated replacement of some rock series by some other ones, the boundaries between such extensive and narrow blocks, being composed of different rocks (for instance, island-arc and shelf ones) showing no indications of transitional rocks. Such blocks are usually separated by faults.

The individual blocks are usually 5-10 km, or sometimes 15 km wide, with their lengths ranging between 100 and 150 km. No correlation can be established between any definite structural and formation zones, except that the sequences of the "island-arc" and "paleooceanic" (siliceous ophiolite) zones are usually restricted to the cores of the synclinoria, as are the Late Paleozoic molasses, whereas the rock sequences with the shelf type (terrigenous-carbonate) rock sequence usually occupy the cores of the anticlinoria or the limbs of the anticlinoria and synclinoria.

Fine folds of varying morphology are distributed extremely nonuniformly in the large folds and in the fold system as a whole. We analyzed this morphological heterogeneity separately for the anticlinoria and the synclinoria to be able to compare the fold dislocations in lithologically similar rocks.

In the South Tien Shan area we studied the structure of seven anticlinoria having sublatitudinal strikes and extending for hundreds of kilometers with the widths ranging between 15 and 30 km. Extending from the north to the south, these are the Kauzan, Kichikalai, and Khodzhaachkan anticlinoria in the Alai segment of the fold system, and the Malguzar, Zeravshan-Turkestan, Zeravshan Range, and Gissar in the Turkestan-Gissar segment.

The anticlinoria are composed mainly of autochthonous carbonate or terrigenous deposits of Middle Paleozoic age and are often arranged en-echelon. For instance, the Kichikalai continues southwest as the Khodzhaachkan, and the latter extends as the Zeravshan-Turkestan.

The folding morphology varies across the strike of the South Tien Shan fold system. The large (first-order) folds composing the cores and limbs of the anticlinoria in the axial part of the fold system (Zeravshan-Turkestan, Khodzhaachkan, and Zeravshan Ranges) are complicated by numerous linear, tightly compressed, often isoclinal smaller folds (of 2nd, 3rd, and 4th orders), which form, along with local faults, upward open fans.

The cross-section structure of these anticlinoria can be termed divergent. There are also gently dipping, bowl-shaped brachyform folds and imbricate fold structures.

Bibliography

Albritton, C. C. : *The Fabric of Geology: Reading,* Mass., Addison-Wesley Publishing Co., 1963.

Babar Md. *: Hydrogeomorphology : Fundamentals Applications and Techniques,* New India Pub, Delhi, 2005.

Belknap, D.F.: *Geoarchaeology in Central Coastal Maine,* New York, Plenum Press, 1995.

Bhagwat, S.B. : *Foundation of Geology,* Global Vision Pub, Delhi, 2009.

Bogli, Alfred: *Karst Hydrology and Physical Speleology,* Springer-Verlag, New York, 1980.

Curran, H. Allen, Young, Drew M., Justus, Philip S., and Garver, John B., Jr.: *Atlas of Landforms,*John Wiley & Sons, New York, 2001.

Davies R. J., Cartwright J., Stewart S. A, Underhill, J. R., Lappin M.: *3D Seismic Technology: Application to the Exploration of Sedimentary Basins,* Geological Society, London, Memoir, 2003.

Duckworth W L H *: Morphology and Anthropology : A Handbook for Students,* Cosmo, Delhi, 2006.

Dunai, T.J.: *Cosmogenic Nucleides,* Cambridge University Press, U.K., 2010.

Edmaier, Bernhard: *Earthsong,* Phaidon Press, London, 2004.

Evans, A.M.: *Ore Geology and Industrial Minerals, An Introduction.,* Blackwell Science, 1993.

Ford, Trevor D. and Cullingford, C. H. D. (editors), 1976, *The Science of Speleology:* Academic Press, New York, 593 p.

Gillespie, C. C.: *Genesis and Geology*: New York, Harper, 1951.

Gurugnanam, B. : *Essentials of Hydrogeology,* New India Pub, Delhi, 2009.

Hallam, A.: *A Revolution in the Earth Sciences,* New York, Oxford University Press, 1973.

Hansen D.M., Cartwright, J.A, and Thomas, D. 2004. The development of intersecting igneous sills from 3D seismic. Geological Society of London, Special Publication. *In:*

Jakucs, Laszlo: *Morphogenetics of Karst Regions: Variants of Karst Evolution,* Halstead Press, John Wiley and Sons, New York, 1977.

Jennings, Joseph N.: *Karst Geomorphology,* Basil Blackwell, Oxford and New York, 1985.

Kondolf, G. Mathias; Hervé Piégay: *Tools in Fluvial Geomorphology,* Wiley, New York, 2003.

Kramer, Steven L.: *Geotechnical Earthquake Engineering,* Prentice-Hall, 1996.

Kuiper, Gerard P. : *The Earth as a Planet,* Chicago, Univ. of Chicago Press, 1954.

Kumar Yadav, Surendra : *Soil Ecology : Remote Sensing and Geographic Information System in Management,* APH, Delhi, 2007.

Mallik, T K : *Marine Geology : A Scenario Around Indian Coasts,* New Academic Pub, Delhi, 2008.

Matt Duckham. : *GIS, a Computing Perspective,* Boca Raton, CRC Press, 2004.

Matthews, M. C. *: Remote Sensing in Civil Engineering,* Glasgow, Surrey University Press; New York, 1985.

McDonnell, R.A. : *Principles of Geographical Information Systems,* Oxford University Press, Oxford, 1998.

Milzow, C.: The Step-pool Morphology of a Steep Mountain Stream, Swiss Federal Institute of Technology, Zurich, Switzerland, 2004.

Nicholls, A.: *Low Attainers and the Teaching of Geography,* Sheffield, Geographical Association, 1988.

Nierenberg, W. A.: *Encyclopedia of Earth System Science,* San Diego, California: Academic Press, 1992.

Okoshi, T. : *Three-dimensional Imaging Techniques,* Academic Press Inc., New York, 1976.

Park, RG: *Foundations of Structural Geology,* Routledge, 2004.

Ramasamy, S. : *Remote Sensing in Water Resources,* Rawat, 2005.

Ritter, D.F., Kochel, R.C., and Miller, J.R.: *Process Geomorphology,* Wm.C. Brown Publishers, Dubuque, Iowa, 1995.

Rondinelli, D.A.: *Applied Methods of Regional Analysis: the Spatial Dimension of Development Policy,* Westview Press, Boulder and London, 1985.

Scheidegger, Adrian E.: *Morphotectonics,* Springer, Berlin, 2004.

Sebba, J.: *Geography for All,* London, David Fulton, 1996.

Sokhi, B.S. and S.M. Rashid : *Remote Sensing of Urban Environment,* Manak, Delhi, 1999.

Strahler, Alan; Strahler Arthur: *Introducing Physical Geography,* Wiley, New York, 2006.

Thorne, C.R.: *Stream Reconnaissance Handbook,* John Wiley and Sons, Ltd., West Sussex, England, 1998.

Trudgill, Stephen: *Solute Processes:* John Wiley and Sons, Chichester, Great Britain, 1986.

Vergne, Jack: *Hard Rock Miner's Handbook.* Tempe/North Bay: McIntosh Engineering, 2003.

Viles, Heather: *Biogeomorphology,* Basil Blackwell, Oxford, 1988.

Wainwright, John; Mulligan, M.: *Environmental Modelling: Finding Simplicity in Complexity,* John Wiley and Sons Ltd, London, 2003.

Williams, Paul W.: *Catena Supplement 25,* Catena Verlag, Cremlingen-Destedt, Germany, 1993.

Yang, Xin-She: *Introductory Mathematics for Earth Scientists,* Dunedin Academic, Edinburgh, 2009.

Index

❑❑❑

www.ingramcontent.com/pod-product-compliance
Ingram Content Group UK Ltd.
Pitfield, Milton Keynes, MK11 3LW, UK
UKHW021405190726
13845UKWH00028B/897

9 789383 263738